Every Root an Anchor

Peck Bur Oak, c. 1837 Wisconsin Historical Society, WHI(C7)6556

When a man plants a tree he plants himself. Every root is an anchor, over which he rests with grateful interest, and becomes sufficiently calm to feel the joy of living. He necessarily makes the acquaintance of the sun and sky. Favorite trees fill his mind, and, while tending them like children, and accepting the benefits they bring, he becomes himself a benefactor.

—John Muir, *Steep Trails*

Every Root an Anchor
Wisconsin's Famous and Historic Trees

Second Edition

R. Bruce Allison
Foreword by Paul DeLong, Wisconsin Chief State Forester

Wisconsin Historical Society Press

Published by the
Wisconsin Historical Society Press

Publications of the Wisconsin Historical Society Press are available at quantity discounts for promotions, fund raising, and educational use. Write to the above address for more information.

Photographs identified with PH, WHi, or WHS are from the Society's collections; address inquiries about such photos to the Visual Materials Archivist at the above address.

Front cover image: Taliesin Elm by B-Wolfgang Hoffmann

Back cover images: Kohler Boxelder and Mrs. Walter Kohler, Sr. (courtesy Julilly Kohler Hausman); Muir Locust (UW–Extension); Twin Lakes Trail Marker Tree

Support for text design and assistance with editing was provided by the Wisconsin Department of Natural Resources, Division of Forestry. An electronic version is available on dnr.wi.gov under forestry.

Printed in the United States of America
Text design by Jeanne Gomoll, Union Street Design, LLC
Cover design by Lisa Wilber

09 08 07 06 05 5 4 3 2 1

Library of Congress Cataloging-in-Publication Data available

∞ The paper used in this publication meets the minimum requirements of the American National Standard for Information Sciences—Permanence of Paper for Printed Library Materials, ANSI Z39.48-1992.

Dedicated to the memory of Walter E. Scott and Robert E. Gard,
two Wisconsin men-of-letters with the character of oak
who chose to spend their lives planting good deeds for the benefit of those to follow.

The Oak

Live thy Life,
Young and old,
Like yon oak,
Bright in spring,
Living gold;

Summer-rich
Then; and then
Autumn-changed,
Soberer-hued
Gold again.

All his leaves
Fall'n at length,
Look, he stands,
Trunk and bough,
Naked strength.

—Alfred Tennyson

He who plants an oak looks forward to future ages, and plants for posterity. Nothing can be less selfish than this. He cannot expect to sit in its shade, or enjoy its shelter; but he exults in the idea that the acorn which he has buried in the earth shall grow up into a lofty pile, and shall keep on flourishing, and increasing, and benefiting mankind, long after he shall have ceased to tread his paternal fields.

— Washington Irving, *Forest Trees*

Contents

Foreword

The state of Wisconsin derives much of its personality from its trees. Whether in the extensive forests that cover nearly half our state, or in Wisconsin communities both large and small, trees are profoundly influential in creating our sense of place. Wisconsin's trees provide our citizens with livelihoods, recreation, spiritual renewal, a vast array of forest products and essential ecological functions that clean our air and water, as well as provide a diversity of plants and animals.

It is no wonder then that, as the chapters in this book relate, trees are woven into the fabric of our state's history and into its people's personal lives. Trees sustain us, and as Wisconsin's first scientist, Increase A. Lapham, warned in his 1867 article "Report on the Disastrous Effects of the Destruction of Forest Trees," we must also sustain them.

Planning a sustainable future for Wisconsin's trees and forests is a primary function of the Wisconsin Department of Natural Resources Division of Forestry. Our mission is to protect and sustainably manage Wisconsin's forest resources so that they provide ecological, economic and social benefits both for us and for the generations that follow.

Enhancing the quality of urban forests, in which 81 percent of us live, is an important component of sustainable forestry. Forests in these urban settings strongly influence the local environment and our quality of life; the benefits trees in our communities provide are many and diverse.

It is in our communities that we also tend to develop special relationships with individual trees. I imagine that as you read this, one or more trees that have been or are special to you come to mind; that's certainly true for me.

We in Wisconsin are blessed with a richness of trees and forests. By collecting and providing these fascinating tree stories in this new edition of *Wisconsin's Famous and Historic Trees,* Bruce Allison reminds us of that blessing and provides further encouragement to manage the resource wisely.

Paul DeLong
Wisconsin Chief State Forester

Preface

I am the warmth of the hearth
on cold winter nights.
I am the shade screening you from
the summer sun.
My fruits and restoring drinks
quench your thirst as you
journey onward.
I am the beam that holds your house;
the door of your homestead;
the bed on which you lie,
and the timber that builds your boat.
I am the handle of your hoe,
the wood of your cradle,
and the shell of your coffin.

— Old European poem

Desolate indeed would be our dwellings were their environs entirely treeless. They are associated with our early recollections and become in a great degree companions of our lives and we unconsciously form strong attachments for such as grow near our homes, thus increasing our love of home and improving our hearts.

— Increase A. Lapham
Wisconsin Oaks, 1856

Trees humanize people. The shade, the warmth, the shelter and the nourishment they provide alleviate our discomforts and allow us to rise to higher, more human planes of thought and action. As a species, our evolutionary roots are in the trees. Our human development was profoundly influenced by tree environments and nurtured by forest resources. Each of us from birth to death is intimately connected with trees. On their beauty and longevity we hang our memories and beliefs, trusting trees to be symbols of our achievements and the things we hold dear.

This book is about both trees and people. Just as the ancients venerated and mythologized trees, from the Norsemen's great world tree, Yggdrasill, to the Greek's oracular oak of Zeus, Wisconsinites, too, have incorporated trees into history and folklore. These tree stories are part of the social history of the state and the personal and emotional history of the people. Affection for trees has influenced our behavior. Trees have served as anchors for time-honored family and community customs, as manifestations of ideals and as reminders of significant events.

People can exhibit great sympathy for trees as was demonstrated by the Brodhead woman who in 1952 offered to pay any price to save a tree from the ax of an uncaring owner. Her efforts rebuked, she immortalized the tree and her cause with a poem. Such loyalty to trees is not uncommon in our state. A farmer in Boscobel, fearing that future owners of his favored oak might not care for it as he did, bequeathed to the tree the land on which it grew. These people saw in their trees more than wood and chlorophyll. Trees to them were companions and friends.

Zona Gale, one of Wisconsin's great writers, explained it this way: "Trees have intelligence. Spirit is combined with them in some degree, in their life and their intelligence. See how they seek out their food, find water, turn to the sun ... there's a better explanation to this than the books make."

Wisconsin architect Frank Lloyd Wright simply stated, "I have mourned the loss of a tree more than a man." Mr. Wright designed his Spring Green home, Taliesin, around a majestic white oak called the Tea Circle Oak. Dignitaries and students from around the world gathered in the shade of this tree to listen and learn from the master. Shortly after Mr. Wright's death, the tree was struck by lightning. Another oak, which had been stunted under the canopy of the original, grew rapidly, rising like a Phoenix to become the new Tea Circle Oak.

Reading through these histories will reveal that trees have served useful purposes in the state, such as witness trees for the early land surveyors. And not so useful purposes, such as hanging trees on which justice was circumvented or, at least, unduly expedited. An example is the Janesville hanging tree on which a mob of incensed citizens strung up an accused murderer. The authorities had the tree cut down to discourage future lynchings.

Other trees marked less infamous historical events, such as the Fort Howard Elm, which was a landmark at the state's first permanent settlement in Green Bay; or the John Muir Locust on the University of Wisconsin campus in Madison under which Muir received his first botany lesson in 1863. A certain historical perspective and cultural humility is gained from trail marker trees. They are living reminders that other cultures have passed this way.

Some trees were turned into green monuments to honor special people, such as the General Douglas MacArthur Pine in Forest County. Other trees, such as the Grant County Sycamore, are growing as memorials to departed loved ones. It was planted by a bereaved father who brought the tree back from Ohio as he was traveling home to Wisconsin with the body of his son, killed in the Civil War. Less tragic expressions of love can also be found in famous trees still growing in the state. The Rhodes Bald Cypress near Kenosha is an unusual species planted over one hundred years ago by a father to celebrate a reconciliation with his daughter, from whom he had become estranged after she married without his consent and moved to a southern state. Just as the Taj Mahal can represent a Shah's boundless expression of love, so can a tree represent the love and devotion of a simple but sincere Wisconsin farmer.

These Wisconsin tree histories are human stories. As Lapham said, they increase our love of home and improve our hearts. They deserve to be told and remembered.

The first edition of *Wisconsin's Famous and Historic Trees* was published in 1982. It was a book idea I had in mind for years. I discovered that others shared the idea. First and foremost was Walter E. Scott, Wisconsin conservationist, author and editor, who for over 40 years gathered information on our state's significant trees. Data from Walter's files gave me a running start on the project. A great deal of information still had to be collected. I naturally turned to our state's great chronicler of historical information, Robert E. Gard. Prof. Gard, who had already invested time on the topic of historic trees, encouraged me and generously assisted by sending a letter to historical societies, the Department of Natural Resources and University Extension personnel, state arborists, newspaper editors and others soliciting relevant tree stories. Dr. Edward Hasselkus, then professor of horticulture at the University of Wisconsin–Madison, also opened his files to me and, as he has done for many others, provided inspiration and direction. I invited Elizabeth Durbin, former editor of the *Wisconsin Academy Review*, to assist with the first edition. She made innumerable phone calls, sent follow-up letters, and made personal visits, bringing the manuscript to completion.

Over two decades have passed since that first edition. Professor Hasselkus and Elizabeth Durbin have retired. My other valued friends and colleagues Walter E. Scott and Robert E. Gard are no longer alive. Likewise, many of the trees alive in 1982 are now gone. These passings make this updated edition even more important to me.

With the encouragement of the Wisconsin Department of Natural Resources Division of Forestry, I revisited each tree history to update it, based on the information currently available. Though I pulled it together, much of the credit belongs to those who shared their stories and information. Some were ambitious collectors of tree data. Others shared specific information about a tree of personal interest. To everyone I extend a sincere thank you for their participation. Names of individuals who were particularly instrumental in the investigation or revision of a tree history are included in the list of sources.

For research and editing assistance with this new edition, I gratefully acknowledge the help of Kathleen McCormick of McCormick Communications and Katherine Esposito of the Division of Forestry, Wisconsin Department of Natural Resources. I especially wish to recognize the skillful editorial assistance and guidance of Kathryn Thompson of the Wisconsin Historical Society Press.

As a professional arborist, my goal is always to preserve trees. Likewise, as a writer, my goal has been to preserve our state's tree-related history. I have received tremendous satisfaction from gathering these histories and saving many faded photographs for posterity.

In setting my goals for the production of this book, I have adopted the trees' time scale, thinking not just of tomorrow or next year, but of 100 years from today when someone will read these stories and gather the fruits of the seed that has been planted. And I hope that, like a tree, the breadth of this project will expand from year to year with new histories told and old ones updated. This is an ongoing project, one in which I hope many people will participate.

R. Bruce Allison
January 1, 2005

Chapter I

Wisconsin's Magnificent Elms

Taliesin Elm B-Wolfgang Hoffmann (1981)

Remembered Elms

The black plague of Dutch elm disease found its way into Wisconsin in the early 1950s. The epidemic spread rapidly from street to street and from forest to forest. The full death count will never be known. In one year alone, the City of Milwaukee lost more than 16,000 trees. Nothing has done more to raise the tree consciousness of the public. "You don't know what you've got 'til it's gone!" And with the passing of the elms we lost a lot. Streets that had inviting canopies of green reverted to barren roadways. Urban parks lost the cooling shade of wide-crowned trees. Home yards no longer had the added architectural punctuation of the elm's graceful arch. Elms, both past and remaining, are a particularly important part of the state's tree history.

A Sawyer County elm, only 15 feet 6 inches in circumference, which blew down in the Flambeau River State Forest in 1952, was 352 years old. It had started growing in 1600, before the first European explorers came to Wisconsin.

There were, and still are, many elms in Wisconsin over 100 years old; some started growing before Wisconsin became a state in 1848. One of the largest American elms was a tree near Gotham in Richland County. It had a circumference of 20 feet 7 inches.

One of the most beautiful elms in Wisconsin stood in a farmyard near Ripon in Fond du Lac County. Its perfectly shaped and symmetrical branches made it a magnificent sight to those passing by.

Elms planted in the 1870s on the University of Wisconsin campus in Madison do not have especially large dimensions, but those trees, as well as elms planted on the state capitol grounds, have been there almost as long as Wisconsin has been a state.

At Lake Mills, in Jefferson County, was a very large elm (now taken down) that probably started to grow in 1821. This tree preceded the first settlers and was doubtless self-planted from a seed which blew there. The first settlers in the Lake Mills area arrived in 1836 when the tree was about 15.

Almost 100 years ago, near Neillsville in Clark County, Celia Reed and Julius Benedict were married. In honor of the occasion, The Rev. William Hendren, who performed the ceremony, twisted together two elm saplings, each smaller in diameter than a pencil, and planted them. In a living symbol of unity, over time they merged together as one tree, eventually towering over 80 feet high. The marriage was also a success, producing five sons and a daughter.

People care about elms. Mount Mary College in Milwaukee had two enormous elms. They were given excellent care, including bracing to protect them from high winds and ice storms. Sister May Seraphia said that "We have no policy regarding the care of old, historic trees except a love of nature, because God gave it to us in his loving providence."

Wellers Elm, Whitewater DNR

General Grant Giant Elm Wisconsin Historical Society, WHi(X3)38161 (1939)

Last elm on State Street, Madison, 1974

Wisconsin Historical Society, WHi(X3)38162

At Whitewater in Walworth County an elm so huge it shaded three lawns stood beside the home of C. H. Wellers. This tree, in 1952, was 96 feet tall and had a crown spread of more than 113 feet. Its circumference was 17 feet 6 inches. Wellers, who estimated that the tree was between 300 and 400 years old, spent a great deal of money in caring for the monarch and protecting it against Dutch elm disease. When he died and the house was sold, such care was no longer possible, and the tree died in the late 1970s. Its loss was noticed and mourned by the entire town.

In Spring Green, in a farm field below Taliesin stood a regal, open-grown elm. It was a landmark leading the way to the home of world-famous architect Frank Lloyd Wright. Regular fungicide injections had protected it against Dutch elm disease. Sadly, the same 1998 windstorm that toppled the Frank Lloyd Wright Tea Circle Oak also caused irreparable damage to the Taliesin elm. It was taken down the next year.

Another huge elm, the General Grant Giant Elm, was located southeast of the intersection of Green Bay and Good Hope Roads, at the site of the Post Village Stage Relay Station, part of the Green Bay–Milwaukee–Chicago Post Mail route.

Sources: Evelyn Cobb, Lake Mills
Mrs. Alfred Kolmos, Whitewater

The Stelter Elm

Marquette County

An extraordinary elm that grew on the Stelter farm near Montello in Marquette County was one of the biggest in the state before it had to be removed, a victim of Dutch elm disease. It measured approximately 15 feet in circumference and had a crown spread of nearly 90 feet.

The Stelters dearly loved that tree. The *Milwaukee Journal* once ran a picture of Theodore Stelter resting under it. The Elm Research Institute recognized the old monarch. Theodore Stelter took the time to write down the following history of their dooryard tree, which had given them so much pleasure for so many years.

Theodore's father, William, bought the farm at auction in 1886, but no tools came with it. William and his father, Gustave, went into Princeton to buy some tools, including a shovel, and on the way home Gustave noticed a small elm tree, about three feet high, growing at the intersection of Highway 23 and Country Trunk C. "You need shade around the house," Gustave said. "Why don't you take the new shovel and dig up that elm and plant it in your front yard."

As the elm grew to maturity, it did shade the house, keeping it pleasantly cool in the summer. The whole family often sat outside on stones placed under its branches. When company came, they took out chairs to sit on. "Even when it was hot, 90 degrees," wrote Theodore, "the wind came under the elm and the air was always cool and comforting. When resting at noon, or when it was very hot, we sat in the shade of the elm to cool off."

After a hospital stay in 1981, Theodore, 75, came back to his home, but he could no longer dress or undress himself. Yet he hated to leave. "The last two nights I slept with shoes and clothes on, unwilling to leave my beautiful home. I cried when I had to leave. We worked very hard and long hours to make it beautiful, so to leave it was heartbreaking."

But the elm tree had gone first. In telling about the removal of the tree in 1979, Theodore Stelter wrote: "It took two big bulldozers to dig under and tip out the stump, about 12 feet by 12 feet. And then it took two big bulldozers to push it into the pasture."

Sources: Theodore Stelter, Montello
J. A. Kronschnabel, Montello

Fort Howard Elm Wisconsin Historical Society, WHi(X313)2141 (1900)

The Fort Howard Elm

Green Bay

It died in the midst of the Great Depression, not in the military surroundings of its heyday, but in the Green Bay yard of the Chicago and Northwestern Railway, inside a protective fence marked with a stone tablet reminding viewers of the large part in history played by the Fort Howard elm.

People had come to view the tree because they could look at the 200-year-old living monument and glimpse, in their mind's eye, what the tree had actually witnessed. Now we have only a photograph, filed in the Wisconsin Historical Society, to remind us of its past.

Undoubtedly, during its military phase, the tree was multilingual. It marked the site of the very first fort in Wisconsin, established by the French about 1718, and witnessed the succession of rule over the territory around Green Bay by France, England and the United States. It loomed over battles of the Fox-Indian Wars, saw the boom in the fur trade, listened in on Pontiac's Conspiracy in 1763, and finally presided over the establishment of the permanent U.S. garrison at Fort Howard in 1816.

Eventually, as settlement progressed, the garrison was dismantled, the land and buildings sold. Opposite the tree, the commanding officer's quarters was torn down. Major Zachary Taylor, who was to become a president of the United States, had stayed there while commandant of the fort. The property became the railroad's, and the fence and the marker went up. And then, there was only the photograph.

The Neenah Treaty Elm

At Neenah, in Riverside Park, at the mouth of the Neenah-Fox River, there used to stand a remarkable elm, the Treaty Elm or Council Tree, so immense in size it was used as a guide by boatmen and steamer pilots on Lake Winnebago. It was also a place where chiefs of neighboring Native American tribes gathered in council.

In 1815, beneath the massive trunk's spreading branches, Winnebago Chief Four Legs began to halt all boatmen on the lake and exact tribute. The Fox Indians had done the same thing a century earlier.

One day in 1819, General Henry Leavenworth, making his way with a convoy of U.S. soldiers up the rapids, encountered Chief Four Legs standing under the tree. The Chief said to the General, "You cannot go through here. The lake is locked!"

The Chief looked imposing in his ceremonial robes, but the General stood up in his boat, lifted his rifle and cried, "So! The lake is locked. Then here is the key in my hand. I will unlock it and go on!" And he took good aim at the Chief, who replied, "Very well. Then you may pass through!"

The Neenah Treaty Elm stood until 1890 when it became necessary to take it down in order to widen the channel of the river. There is, however, a photograph of this old elm in the Wisconsin Historical Society at Madison, and a slab from its wood has been made into a table top in Governor Doty's old cabin at Neenah, preserved as a relic of pioneer times.

Neenah Treaty Elm

Wisconsin Historical Society, WHi(X28)2142 (1908)

The Grignon Elms

Kaukauna

The stately elms planted around the Grignon mansion in Kaukauna in Outagamie County were, until they fell victim to Dutch elm disease, majestic symbols of one of Wisconsin's earliest and most important families.

The 12-room, three-story house, open to visitors now, was built in the woods about 1837 by Charles A. Grignon, and provided a home for his descendants for almost 100 years. Until the 1970s, many of the elms planted by Charles A. were still standing. The largest one had a crown spread of 140 feet and a circumference of 21 feet 8 inches when it was removed in 1973, another casualty of Dutch elm disease.

Charles A. was the grandson of Pierre Grignon who established Wisconsin's first trading post on the site in 1760. Descended from French nobility, Pierre married Domitille de Langlade in Green Bay in 1773. Domitille was the daughter of Charles de Langlade, one of the founders of that first permanent white settlement in Wisconsin. Twenty years after the wedding, Pierre's trading post site in Kaukauna was purchased by Dominique Ducharme from the Native Americans for two barrels of rum. The deed legalizing this sale is the oldest such document in the state.

The property, with the Ducharme log home on it, later returned to Grignon hands when Ducharme sold it to Augustin, son of Pierre and Domitille. When it passed to Augustin's son, Charles A., he constructed the elegant Colonial mansion and planted the elms.

The last of the Grignon family, Mrs. Ross Grignon, moved out of the home in 1935, when the old elms were still casting their shade over the spacious grounds along the banks of the Fox River. The property became a historic site shortly afterwards.

Source: Victor Luedtke, Kaukauna

Grignon Elms

Wisconsin Historical Society, WHi(X3)1947 (c. 1950)

Indian Agency House and Portage Elms

A very old elm still stands guard near the front entrance of the Indian Agency House off Highway 33 in Portage. The U.S. Government built the house in 1832 for John H. Kinzie, Indian Agent at Fort Winnebago, and his wife, Juliette Magill Kinzie.

Years earlier, John Kinzie's family moved from Sandwich, Upper Canada, to Fort Dearborn in Chicago, where they survived the massacre of 1812. With the assistance of their Indian friends, they fled to Detroit. In Michigan, John learned the fur trade and eventually went to work for the American Fur Company headquarters in Prairie de Chien. Because of his extensive knowledge of Indian customs and languages, John often served as interpreter, including with the Territorial Governor, Lewis Cass, who later appointed John as the Indian Agent to the Ho-Chunk (then called Winnebagos).

John and Juliette Magill were married in Detroit in 1829. They traveled by steamer from Detroit to Green Bay. From there, they traveled by Mackinac boat up the Fox River to Fort Winnebago. Juliette brought all of her belongings with her, including mahogany furniture, silver, china, and a large piano.

When John and Juliette arrived at Fort Winnebago, their house was not ready for them. They stayed in the Fort, where Juliette and her piano were the wonder and joy of both the soldiers and the Native Americans. The Agency House was completed in 1832, and the Kinzies moved in. The dooryard elm, now very old, is still an impressive tree. The Kinzies also planted maple trees for shade, but those trees have since died.

About a mile east of where the maples were planted stood a solitary elm left from a group of three in front of the first tavern/hotel in that part of the country. Situated on the Wauna Trail—the route between the Fox River and the Wisconsin River used by portagers—the Franklin House, as it was called, was owned and operated by Captain Gideon Low. Beginning August 30, 1847, with David Irvin on the bench, the first court in Columbia County was held there. The tavern is long gone, and the tree was cut in January 1963 because of a large amount of decay.

In 1902 the Portage (Waubun) Chapter of the Daughters of the American Revolution planted elm trees to mark three historic sites: the beginning of the Wauna Trail on the banks of the Fox River, where the Native Americans and *courier du bois* drew their canoes ashore to transport them across the short stretch of land to the Wisconsin River; the end of the trail at the Wisconsin River; and the cemetery of the areas's first Catholic Church where Pierre Pauquette, the famous Portage strongman, Winnebago interpreter and carrier of canoes, was buried. These elms are all down. A fourth elm was later planted at the corner of West Cook and MacFarlane Streets where the popular and well-liked Pauquette, a man so strong he could lift a horse, was murdered after a quarrel in 1836.

Sources: Mrs. Maynard Benson, Portage
Rita Fredrick, Portage
Dorothy McCarthy, Portage
Joyce Steward, Madison

Milwaukee's Menomonee River Parkway Elm

At one time the state record American elm towered over its fellows in Milwaukee County's Menomonee River Parkway, near Center Street in Wauwatosa. It had reached the majestic height of 145 feet and had a circumference of 21 feet 2 inches before it died in 1968, a victim of Dutch elm disease. Its age was 296 years, which meant it had established itself just a few years after the arrival of Jean Nicolet, the first French explorer to set foot in the territory, which became Wisconsin.

The tree had been in perfect health up until a few months before it died. It had been sprayed with DDT and injected with Bidrin every year to protect it from the bark beetles that carry Dutch elm disease, and the county forester had planted densely growing shrubs and thorny bushes around its base—and even a little poison ivy—to discourage vandals.

The tree missed its annual protective treatment in 1968 because high water prevented the county forester from reaching it. The elm was in a hollow some 75 yards off the road, near the river.

"It was a very beautiful tree," the forester lamented in a *Milwaukee Journal* article. "We're very upset and we're going to miss it. It will leave a big gap." He said he would save some cross sections of the trunk. "It's the only thing we'll have to remember it by."

Menomonee River Parkway Elm *Milwaukee Journal* (1968)

Tuesday Club Elm (1965)

The Tuesday Club's Slippery Elm

River Falls

Perhaps the only tree in Wisconsin ever protected by a woman's club was a slippery elm *(Ulmus rubra)* located in the city of River Falls in Pierce County. Eighty feet tall, it was estimated to be well over 100 years old and had a trunk circumference of 16 feet 5 inches.

The Federated Women's Club of River Falls, known as the Tuesday Club, launched a campaign to acquaint the community with the problems of Dutch elm disease. Their work in this area was so outstanding that an award of $200 was given to them by their state organization. This money was used, with additional funds coming from the city, to preserve the landmark slippery elm as long as possible. A large crack that had developed was filled and the tree was reinforced with cables and bolts. Unfortunately, it too succumbed to Dutch elm disease and no longer graces the small park overlooking the Kinnickinnic River.

Source: Catharine Lieneman, River Falls

The Rockton Grove

A letter from Joyce McVey Blackmore, June 12, 1981, reads:

There was a beautiful stand of stately elm trees in the village [of Rockton, Wisconsin] called "The Grove." I have no idea how many trees were there at one time, but there were many.

These trees were on land owned by Van S. Bennett, who ... was the largest landed proprietor in Vernon County, owning about 2300 acres. He was a member of the State Assembly in 1869–70 and of the State Senate in 1884 to 1886.

He laid out and platted the village of Rockton in 1873, donated land for the cemetery and for the school, and was responsible for having the road built across "The Island" where the Kickapoo River was divided, built a large mercantile store, and was responsible for many other things.

Two men of the Rockton community—Christian Selback and Eli McVey—built a "Swing" out of wood in "The Grove." A horse pulled the swing around, and children rode it for 5¢ a ride.

The first large gathering of people ever assembled in this part of the county was on the 4th of July, 1868, in "The Grove." One old lady shed tears of joy to see so many people gathered together "in the woods." Hartwell Allen, the first county school superintendent, was the day's orator.

On August 13, 1939, Rockton's first homecoming picnic was held amongst these trees. A very large crowd attended. There was a program of speeches and music. People visited, played horseshoes, and watched Oscar Sayles' team of oxen perform.

It was all so successful that it was decided to make it an annual affair on the second Sunday in August. Later it was held on the lawn of the old "Brokaw House" across the road where many of the same elm trees were.

Eventually, in the late 1950s, the annual homecoming was discontinued because there were no longer enough of the older residents of Rockton left to do all the work necessary.

Also, by this time, many of the famous old elms had succumbed to lightning, wind, or disease. Many are still left, though, so one can see their beauty and realize their importance to this pioneer community.

Source: Joyce McVey Blackmore, Phoenix, AZ

Prince Philip plants elm developed by Eugene Smalley (right)

Fox Photos LTD. (1980)

Disease Resistant Elms

A Wisconsin elm hobnobs with royalty on the grounds of Windsor Castle in England. Prince Philip, husband of Queen Elizabeth, planted the young tree In April 1980, using a silver-plated shovel, while its developer, Professor Eugene B. Smalley, of the University of Wisconsin–Madison, stood proudly by. The "Elms Across Europe" project was launched in 1979 by Pitney Bowes, an international business machine firm.

This elm, sometimes called a super elm, was gaining worldwide attention as the first commercially developed elm resistant to Dutch elm disease.

Since it was discovered in the Netherlands in 1929, Dutch elm disease has killed well over 50 million elm trees on both sides of the Atlantic. By 1980, Britain had lost more elms than most European countries, and the beauty of its landscape suffered accordingly. An elm tree that would resist the disease had long been sought.

When Professor Smalley joined the UW–Madison faculty in 1957 as a plant pathologist and mycologist, his mission was clear: to find a solution to Dutch elm disease. According to forest ecology and management professor Ray Guries, "One of the first things Gene did was to begin assembling genetic resources here in Wisconsin." Smalley began planting seedlings at the 50-acre Arlington Experimental Farm north of Madison and on his own 20-acre property near Cottage Grove, informally known as "Smalley's Mountain Research Sanctuary."

After 20 years of collecting and testing elm seedlings from all over the world—including Europe, China, Canada, and other parts of the U.S.—Professor Smalley developed a hybrid that would be generally disease resistant, cold hardy, and immune to Dutch elm disease. He named his first hybrid Sapporo

Autumn Gold because it grew from open-pollinated seeds collected from a Siberian elm growing in the Botanical Garden at Hokkaido University in Sapporo, Japan. The Hokkaido seeds had resulted from the natural hybridization of the Siberian elm, which was disease resistant, with a nearby Japanese elm, which had a pleasing shape. Other hybrids soon followed, including Regal, American Liberty Elm Independence, New Horizon, and Cathedral.

Smalley's disease resistant elms have been planted all over the world, including along the autobahns and Rhine River in Germany, in London's Hyde Park and at Windsor Castle. They have also replaced dead elms on the UW–Madison campus, Wisconsin State Capitol Park, at Harvard University, and along streets in cities all across the U.S.

Professor Smalley died at the age of 75 in March 2002. Professor Guries described Smalley as someone "incredibly loyal to the university and the state of Wisconsin. Once Gene came and started the program, he stayed," he said. "He came and he stayed because the trees were rooted here. And, in a funny kind of way, he became rooted here, too."

Sources: Dr. Eugene B. Smalley, Madison
Dr. Raymond Guries, Madison

Chapter II

Historic Trees

The trunk of the original Frank Lloyd Wright Tea Circle Oak can be seen just behind the second Tea Circle Oak. Frank Lloyd Wright Foundation

The Frank Lloyd Wright Tea Circle Oaks

Spring Green

Frank Lloyd Wright, one of the world's great architects, felt the natural surroundings of a home or building should be integrated into the overall design of the structure. Though not physically tall, Wright was a towering figure in his genius, and perhaps he identified with a mighty white oak growing on the site he chose for his own home near Spring Green in Sauk County. Some have called Taliesin—the sprawling complex that includes workshops, training ground, farm, recreational haven and living quarters—Wright's finest achievement. He constructed the Tea Circle around two old oaks. There, on warm afternoons while tea was served, Wright would converse with the students who had come to Taliesin to live, work and learn from the master. This practice continues today with the present Fellowship members, even though Wright is gone.

Wright died at 90 in April 1959 at his Arizona winter home and headquarters. Back in Wisconsin soon afterwards, a bolt of lightning shattered the larger Tea Circle Oak beyond saving. However, the smaller oak that had started life under the spreading canopy of the larger tree quickly filled out to replace the Tea Circle Oak that had been hit. The second oak thrived, its canopy eventually extending outward in a diameter of over 100 feet, providing shade and inspiration for the next generation of architects at Taliesin. Tragically, the second original Tea Circle Oak was blown over by a devastating windstorm in 1998. With straight-line winds of over 80 miles per hour, the storm toppled about 15 trees on the estate, including the remaining Tea Circle Oak, which snapped at the base and fell onto the roof above Mr. Wright's studio. R. Bruce Allison and his tree company crew had the daunting task of removing it limb by limb from the building. Many of the remaining original Wright understudies at Taliesin mourned the loss of the tree. Then, with characteristic perseverance and determination, they arranged to have a new oak

Second Tea Circle Oak Frank Lloyd Wright Foundation

planted to carry on the tradition. As a symbolic and actual replacement, it seems to embody the spirit of the place, which Wright determinedly rebuilt twice after disastrous fires.

During almost 50 years as master of Taliesin, Wright was often photographed under the older oak with visiting dignitaries who came to pay homage to the greatness of his original architectural achievements, or to visit the magnificent home itself. The name *Taliesin* is Welsh and means *shining brow,* and indeed the house, integrated in organic wholeness with the landscape, curves like a brow on the side of the hill. Wright made its prairie house style famous, and it truly belongs among the rolling green hills and oak openings of southern Wisconsin.

It is natural to compare Wright to an oak. A man of great strength and will, he survived devastating personal tragedies, an often indifferent or hostile public, and a challenging press. However, he is now generally recognized as a creative genius. He designed more than 600 buildings, pioneered a completely new concept called organic architecture and developed numerous innovations that would be followed for years to come. Perhaps only a man as strong as an oak could withstand the setbacks he endured while never losing the confidence in his own ideas and abilities that enabled him to forge ahead in spite of indifference or criticism.

Sources: Aubrey Banks, Arizona, Spring Green
John Hill, Arizona, Spring Green
Doug Hadley, Taliesin

Second Tea Circle Oak, fallen in 1998 Taliesen Preservation, Inc.

Dunbar Oak — Waukesha County Museum Collections

The Dunbar Oak

Waukesha

Desperately ill and convinced his death was imminent, Col. Richard Dunbar sank down to rest under a strong white oak in Waukesha County one August day in 1868. A diabetic suffering from insatiable thirst, he had just drunk six tumblersful of water from a spring beside the tree. He began to feel better. He continued to drink freely from the spring waters, and soon the man doctors said would die within months felt completely cured. Dunbar lived ten more years without a recurrence of his diabetes, and during this decade he made Waukesha, the spring waters and the "guardian angel" oak tree world-famous.

His faith in his own cure inspired him to offer the waters to others; his sharp business ability made it financially rewarding. The former engineer bought the spring, which he named "Bethesda," and the land around it and actively began to promote the community. Waukesha soon became a haven for those seeking not only restoration of their health but also relaxation. Soon Dunbar was selling the spring waters all over the globe.

Other promoters entered the picture and several more springs were developed, among them Hygeia, Arcadian and White Rock. To quote from an account of those times by Libbie Nolan, of the Waukesha County Historical Society:

> *Fancy ornate pavilions enclosed each bubbling spring ... Great hotels sprang up.... Guests came from far and wide to drink from the springs, to relax, and to be amused. Among them were Mrs. Abraham Lincoln, President Grant, Vice President Adlai Stevenson, Supreme Court Justice Salmon P. Chase, Horace Greeley as well as governors, generals, statesmen and many many more.*
>
> *White Rock water became a fashionable beverage all over the world after it was carbonated in 1890, even making its way to the coronation of the King of England ... When Col. Dunbar died in 1878 (of a heart attack), the village was notified by the passing out of black-edged handbills.*
>
> *The fame and excitement of our Western Saratoga lasted nearly forty years. Then it faded away as the century turned.*

Dubbed the Dunbar Oak, it was a focal point of the large park at Bethesda Spring and is marked by a plaque put up by the Waukesha County Historical Society in 1959. After noting Dunbar's first introduction to the waters, it ends with the words: "The names 'Dunbar,' 'Bethesda,' and 'Waukesha' became household words. Thus began the glamorous and exciting summer resort era of Waukesha as the 'Saratoga of the West' 1870–1905."

The Dunbar Oak went down during a severe thunderstorm in 1991. Prior to that, the oak had been cabled because it had developed a pocket of decay that made it vulnerable. But with winds clocked at close to 100 miles per hour at the Waukesha airport, the trunk of the Dunbar Oak snapped at about 10 feet above ground. A count of the annual rings revealed the tree was over 320 years old. The loss of the tree affected many locals, as evidenced by the long lines of cars that paraded past the site where it fell, as if people wanted to pay their last respects. Waukesha City Forester David Liska, planning for the future, took cuttings from trunk sprouts to Johnson's Nursery. There they grew in secret for 12 years. In May 2004, in celebration of Waukesha's 25th anniversary as a Tree City USA and Wisconsin Forestry's centennial year, a "reborn" Dunbar Oak was planted at the site of the original and watered with 15 gallons of donated Bethesda Springs water.

Sources: Pat Buckley, Waukesha
Jean Penn Loerke, Waukesha
David Liska, Waukesha

Trading Post Oak

Lake Mendota

A big red oak marked one of the most historical sites in Dane County on the northwest shore of Lake Mendota where an early trading post once stood. It is most famous as the place where Colonel Dodge, accompanied by Colonel Gratiot, spoke to more than 5,000 Winnebago Indians in 1832. Runners, sent out by Dodge, had brought the Indians together during the crisis of the Black Hawk Rebellion. Dodge exhorted the Winnebago to keep the peace as the Sauk war leader Black Hawk and his 1,000 or so followers—men, women and children—retreated through the area on the Sauk Trail after a futile attempt to regain their lands on the Rock River. Commemorating the spot, along with the oak, is a granite boulder. The inscription on its bronze plaque reads:

Trading Post Oak Robert Keller (c. 1980)

Site of the Rowan-St Cyr fur-trade post
1832-1837
Site of city of Four Lakes
Colonel Henry Dodge met
Winnebago Indians here
In council May 25, 1832

The location is now 3119 Waconia Lane in Middleton.

Trader Wallis Rowan originally built the shack, then sold it to trader Michel St. Cyr, son of a French Canadian father and a Winnebago mother. St. Cyr, noted for his honesty and fair dealing, made many improvements. Because there were so few dwellings in the county, the trading post often served as an informal hotel, with travelers spending the night for $2, a fee that did not make St. Cyr rich.

Among the overnight guests at the St. Cyr cabin were Moses Strong, who platted the city of Madison, and his party. They enjoyed their host's specialty, muskrat pie. James Doty, territorial governor, stayed there also and built a hunting lodge on the property. The hunting lodge still stands. So scattered were habitations that in 1837 St. Cyr's overnighters included two men who were famished after traveling more than 36 hours without food except for a few cold potatoes found in an unoccupied Indian wigwam. St. Cyr's muskrat stew must have tasted mighty good to them.

In 1834, Col. William B. Slaughter, register of the land office in Green Bay and also involved in the organization of Wisconsin Territory out of Michigan Territory, arranged a deal to register St. Cyr's lot. Slaughter paid the trader $200 for the improvements he had made and became an owner. St. Cyr subsequently conveyed an interest to James Doty as well. Slaughter, Doty and Morgan L. Martin of Green Bay then platted out the city of Four Lakes around the red oak and the trading post, hoping it would be chosen as the capital of the new state. Like Belmont, Four Lakes came close, but missed out. Doty had, of course, ensured his own success by selling Madison property to several territorial legislators as well.

Reportedly St. Cyr, a carpenter as well as trader and hosteler, helped to build the first post office in Madison, a log edifice located on the Capitol Square. He eventually moved on to Minnesota where he died in 1859.

Source: E. G. Kromrey, Middleton

Peck Bur Oak, c. 1837, from an 1869 painting by E. E. Bailey

Wisconsin Historical Society, WHi(C7)6556

The Peck Bur Oak

Madison

The bur oak tree beside the Peck cabin in Madison witnessed an impressive list of pioneer firsts: first residence in the fledgling settlement, first tavern, restaurant, hotel and post office, site of the first wedding and birthplace of the first white child born in the capital city.

Eben and Rosaline Peck had decided to move from the Brigham farm in Blue Mounds on the advice of Judge James E. Doty, who had reason to be optimistic about Madison's potential for growth. Eben arrived first to start construction of the cabin during the winter of 1837 on land bought as soon as the territorial legislature, nudged by Doty, chose Madison as the location of the state capitol. Eben's pregnant wife followed, as described in *The History of Madison, Dane County and Surroundings*, published in 1877: "Snow and the howling of wolves awakened Mrs. Peck from her slumbers in a tent, three miles from Madison, on Saturday, April 15, 1837, and [with her four-year-old son Victor] she pushed on through the storm to the site of her more substantial dwelling, where she sat down under a tree in her wagon, twenty-five miles from the nearest white residents at Blue Mounds and nearly one hundred miles from the settlers at Milwaukee. The building was not far enough advanced to satisfy the demands of the hostess and a temporary habitation was constructed...."

In the blizzardly desolation of her arrival with only a bed quilt to throw over her own and Victor's head, Rosaline must at least have appreciated the presence of that stalwart oak!

With the cabin finished a few weeks later, Judge Doty, ever the helpful guest, spent a day with "all the available masculinity of Madison" plastering Mrs. Peck's kitchen in order to ready the premises for boarders. Once plastered, the "little hotel was speedily crowded with guests." As *The History* relates, "About two weeks after the arrival of Mrs. Peck in Madison, a party of 15 men came on from Milwaukee via Janesville, and the work of the hostess began in earnest." Besides Judge Doty, Colonel Ebenezer Brigham, first settler in Dane County; Commissioner Augustus A. Bird, one of the three men elected to supervise the construction of the capitol; and other "historic personages" were frequent visitors. A traveler from England, one George Featherstonehaugh, when safely back in London, wrote that Mrs. Peck's coffee was not good. No less a luminary than General Simeon Mills, a prominent early Madison citizen, disputed this slander, commending Mrs. Peck's "excellent coffee." In fact, Mrs. Peck reportedly set a table that was "a marvel to beholders" and "cleanliness, the first requisite towards elegance, was a welcome feature from the beginning."

July 4, 1837, marked a historic occasion: the laying of the capitol cornerstone. Although others dispute the numbers, Mrs. Peck described a celebration of the event that went on for several days and involved two to three hundred persons, including the friendly Winnebago Indians living a short distance away on the shores of Lake Monona.

A year later the Peck "tavern stand," as Mrs. Peck phrased it, came under the proprietorship of R. L. Ream, father of the famous sculptor Vinnie Ream, when the Pecks went into full time farming. Under Ream's management, the Madison House "charges were not very moderate, as 'two feet by six of floor could be had for the night,' only upon payment of 'two pence per square foot,' and 'the weary traveler might spread his own blanket, using his saddle or portmanteau for a pillow, rejoicing that he had so good a bed.'"

The Peck cabin, after 20 years as the Madison House, was razed in 1857. The oak stood until 1930, when workmen preparing the site for the Farm Utility Building cut it down. The state office building GEF III now occupies that spot with a plaque noting the site's history.

With all other traces of that early history being so completely obliterated, it is pleasant to recall the impact that Eben and Rosaline Peck had on the community, as summarized in the above-quoted *History*: "In her house the earliest visitors to Madison found a home, in her dining room the gayeties of several seasons found their earliest expression. Her husband as justice of the peace united in the bonds of wedlock the first couple lawfully married in this city, and after the irrevocable knot had been tied ... the violin of the justice's lady gladdened the hearts of the assembled throng while they threaded the mazes of the dance. In the old log house was born Miss Wisconsiniana Victoria Peck, the first child that saw the light of this city... Mrs. Peck and her husband were the pioneer settlers...."

Source: Frank Custer

The Plough Inn Maple

Madison

A fascinating history of the early days of Wisconsin could consist only of tales about taverns. Often the first residences in an area, these sometimes rough and ready hostelries sprang up along the military, territorial and plank roads, especially on stagecoach routes. They supplied food and drink, entertainment, warmth and society, and, if not a bed, some floor space and a buffalo robe to the haulers, teamsters, stage coach passengers, foot travelers, wagoneers and horse back riders who thronged those early thoroughfares. The Old Plough Inn on Wiota Road was such a place.

Now an ecologically friendly bed and breakfast called Arbor House, it still stands at 3402 Monroe Street in Madison. It is marked with a plaque, put up in 1973 by the Daughters of the American Revolution (DAR), stating that the Plough Inn and stage coach stop is one of the oldest existing houses in Madison, built in 1853 on a tract of land entered in the U.S. Land Office in 1836.

Until 1981 when a fierce windstorm demolished it beyond repair, one of the oldest maple trees in Wisconsin looked over the structure. It was there long before the land was platted or passing travelers needed a spot to rest and refresh themselves.

A stonecutter named Paunack built the home/inn of brick and native sandstone from a nearby quarry, and its 18-inch walls, hand-hewn beams and wide plank maple floors attest to its strong construction.

John Whare, a short, heavy Englishman with a quick temper, bought the place in 1858 and turned it into a tavern. He also sold plows (being an Englishman, he spelled it "plough"), hence the earlier name. On the lower floor he served soup and beer to miners from Mineral Point and wagoneers hauling loads of ore. Dancers could whirl and stomp on the upper floor. Always ready for a fight himself, Whare was well able to handle any rowdiness that might occur under his roof. Apparently there was considerable, much to the detriment of the furnishings.

Harry Ellsworth Cole, in his book *Stage Coach and Tavern Tales of the Old Northwest*, wrote that Whare was as testy with his family as he was with his unruly guests. One day at dinner, when his son angered him, he knocked the boy clean across the room. Recovering quickly, the son dumped a steaming bowl of

Plough Inn Maple Wisconsin Historical Society, WHi(X3)28642

gravy over his parent's head and sprinted for the woods handily surrounding the tavern, as his Pa grabbed for the shotgun. How long the youth hid out is unknown.

Another tale passed along from guest to guest, probably in front of a flickering fire, on a dark and stormy night, claimed that the body of a man killed in a drunken brawl at the Plough Inn lay buried a short distance away. There are similar tales about other taverns, and maybe they are all true. In those days, neither accommodations nor patrons were always elegant, and frontiersmen often made their own laws.

Source: Lorraine Wilke, Madison

The Spring Tavern/ Hotel Walnut

Madison

It's hard to remember that well-established Madison neighborhoods were once considered suburban or even way out in the country. Nakoma is one of them, and the old Spring Tavern/Hotel at 3706 Nakoma Road, once well outside city limits, is famed as the first hostelry between Madison and Monroe.

An old black walnut growing beside it since Indian times shades the front lawn in summer and stands stalwart and imposing during winter storms.

Native Americans constructed burial mounds on the property. Archeologists found skeletons in a sitting position, along with pots that had contained food, grinding stones, battle axes and copper knives. These items were transferred to the Wisconsin Historical Society, then located in the basement of the first capitol, and were unfortunately lost when the building burned in 1904.

Spring Tavern Walnut

B-Wolfgang Hoffmann (1982)

Nakoma Road wasn't there, of course, when the Spring Hotel had its heyday. The road that did lead out of Madison along the route of Monroe Street has been variously called, depending on your source, the Wiota Road, the Military Road, the Pioneer Road or the Black Hawk Trail. Whatever its designation, it was heavily traveled by the wagoneers and haulers serving the lead mines of southwestern Wisconsin. The stagecoach made regular stops at the Spring Hotel, while mail carriers rendezvoused there.

The free-flowing spring (now known as the Arboretum's duck pond), which gave rise to the name, attracted dusty travelers, served as water supply and hosted wild fowl year round. Because the land sloped down to the spring, the 15-room structure was two stories in front and three stories in back. Later a handsome Georgian Colonial porch with square wood piers and a latticed balustrade was added on the Nakoma Road side.

Charles Morgan, who came from New Haven, Connecticut, built the spacious home in 1854. Morgan made the small, vermillion-colored brick from clay found on the slope with the help of his neighbor, a man named Gorham. The oaks that undoubtedly stood on the property were probably cut down to make the beams and planks for the house as well as others farther up the hill. There are no oaks left.

James Gorham bought the house and 60 acres in 1860 for $9,000 in gold when Morgan became a dry goods merchant in Madison. The Gorhams owned it for some 60 years.

The hotel became, besides a haven for travelers, the last overnight before Madison, a social center famous for its dances, to which soldiers stationed at Camp Randall escorted female students from the University. Supposedly three presidents and quite probably Robert M. La Follette visited the place, and it became famous for its cookies—not Toll House cookies as the city inscription in front of the house alleges (Toll House cookies were invented in 1930 by Ruth Wakefield, who ran the Toll House Inn in Whitman, Mass.), but probably a large sugar and butter cookie studded with raisins.

Professor James Dickson bought the house in 1923, greatly improving and restoring it, and in 1939, William and Florence Stephens took over the charge. They maintained the home in mint condition, adding greatly to the beauty of the landscaping and gardens, and were instrumental in having the home listed in the National Register of Historic Places. In 1972, the John Bell Chapter of the DAR put up a plaque on the Frank Lloyd Wright–designed stone wall along the Nakoma Road side of the house.

Sources: Edward Hasselkus
Mr. and Mrs. William Stephens

The La Follette Hickory

Madison

Everyone calls it the La Follette Hickory, but no one knows why. It is not a large tree, but then it grew for many years in the shade of the graceful elms that made such a compatible setting for the impressive state capitol in Madison.

Richard Smith, a former groundskeeper at the Capitol Park, remembers hearing about the La Follette Hickory from his boss. "It was a little thing when I first came to work there in 1949," he says, "and it must have been planted sometime in the '30s but it wasn't growing. But then we trimmed it up and gave it some fertilizer and finally it started to grow." Smith didn't remember why it was called the La Follette Hickory.

La Follette Hickory B-Wolfgang Hoffmann (1982)

The capitol tour guides still call it the La Follette Hickory, quickly adding that they don't know why. It may be rumor or folklore, but the tree could have been planted during the tenure of Philip La Follette, Fighting Bob's younger son, who won the governorship from Walter J. Kohler in 1930 when he was only 33 years old and served six turbulent years attempting to make the state a national model for solving the problems of the Depression.

In the words of Robert C. Nesbit, author of *Wisconsin, A History*, "Phil was the son who had inherited Old Bob's political sense, oratorical ability, and zest for the campaign." But inevitably, Wisconsin leadership could not resolve national crises and the focus shifted to the policies of Roosevelt and the federal government. Philip F. La Follette dominated Wisconsin politics throughout the 1930s, but his "ambitions and frustrations carried him out of the mainstream to political oblivion," as the Progressive Party enjoyed its last decade of real influence.

The La Follette Hickory, on the other hand, has thrived.

Sources: Stan Cravens, Madison
Leon Hendrikson, Madison
Jack Holzhueter, Madison
Robert Shaw, Madison
John Short, Madison
Richard Smith, Madison
Mike Stark, Madison

The Coffin Tree

Rusk County

A tale worthy of a Halloween telling involves a large, hollow basswood tree, a forest near Ladysmith in Rusk County and two loggers, Art Charpin and Walter Latsch, who were scared out of their wits on a February day in 1926. It was first told by a Rusk County weekly.

The day started out innocently enough. The basswood had been marked by a timber cruiser for cutting and Charpin and Latsch were wielding the saw. They noticed a large hole in the tree some 30 feet above the ground, but they felled it anyway, figuring they'd get a 20-foot log out of it. The trouble started when they tried to saw through the trunk. The saw bound up against something hard. They came in from another angle. The saw again hit a rocklike center. At this point they were curious about a rock inside a tree, and turning the log as needed with a cant hook, they managed to saw all the way around the "rock" and pull the end of the trunk away. The wrinkled face of a man stared up at them. Shaking, they hightailed it back to town.

Eventually their story was believed and a party of four went into the woods to investigate. Sure enough, when the trunk was completely removed, they found the body of a man, fully clothed in coarse homespuns and buckskin, which fell away when touched. The head, covered with long hair, still wore a coonskin cap. With the mummified body the men found an old muzzle-loading flintlock rifle and a fancy muzzle-loading pistol. Pieces of paper found on the body at first seemed to identify the man as a Captain D' Artagon who had been with the Marquette-Joliet party seeking a route to the Pacific Ocean in 1673. This seemed totally implausible, however, as the exploring party never came near Rusk County, and if the man had been lost, it is not likely that he would have continued to travel westward, away from the Wisconsin River.

The finders of the grisly surprise finally theorized that the man, whoever he was, had been pursued by Indians, had taken refuge in the hollow tree, and unable to get out again, had died there.

According to the Rusk County weekly, the body was supposedly shipped to the University of Wisconsin–Madison. The two loggers vowed never to return to the haunted forest near Ladysmith.

The Knut Juve Trees

Dane County

On the farm of Knut Juve in Pleasant Springs Township in Dane County stood two oak trees under which one of the earliest Norwegian congregations in Wisconsin gathered on September 2, 1844. The pastor was the Reverend J. W. C. Dietrichson, newly arrived from Norway, and soon the congregation dedicated a new church, a log one, which is believed to be the first Norwegian Lutheran Church consecrated in America.

The last of the two oaks toppled in a windstorm in 1923. Local residents have some mementos such as letter openers made from this tree. The land where the trees stood is still in the Juve family. A granite marker indicates the significance of the trees:

This memorial marks the site
of the first sermon preached here
by the Rev. J. W. C. Dietrichson
Sept. 2, 1844
West Koshkonong Ev. Lutheran Church

Boscobel's Dean Oak

If fame means anything to a tree, the 300-year-old Dean Oak near Boscobel in Grant County should feel well honored. Sometimes called the Charter Oak, the venerable monarch has had its share of recognition. It has had its portrait painted in oils and it is, by verbal agreement at least, a landholder.

The painting of the white oak, by R. F. Ferris, hangs in the Boscobel Public Library. As for its landowner status, it may not be a matter of record in the Grant County Courthouse, but old-timers assure us of its truth.

The tree was large and impressive when Charles K. Dean, the founder of the city of Boscobel, bought the land on which the tree stands. He grew exceedingly fond of it. When, prior to 1890, a 12-acre piece of the Dean farm was sold to John Verrill, pioneer Boscobel horse trader, Dean made the oral stipulation that the tree was to have one-third of an acre deeded to it and was never to be cut during its lifetime.

Undoubtedly, the old oak has witnessed much history. It grows one mile east of Boscobel on Homer Road (County Trunk S), beside what used to be an Indian trail connecting Galena, Illinois, the end of the railroad from where so many settlers walked into Wisconsin, with Prairie du Chien. One can imagine parties of pioneers or Native Americans stopping beneath the tree's pleasant shade to talk, play a game or rest. In fact, the oak is said to serve as the final resting place of an Indian chief known as Thunderbird.

Today the Dean Oak is not as majestic as it once was. It has lost some of its branches to lightning, and windstorms have taken their toll.

Source: Ralph Goldsmith, Boscobel

Boscobel's Dean Oak DNR (c.1955)

Boscobel's Dean Oak B-Wolfgang Hoffman (1982)

The Cooksville Scots Pine

A record Scots pine kept the peace in an old cemetery in Cooksville, Rock County, a cemetery that might not have been there if Daniel Webster, the famous United States Senator, hadn't speculated in Wisconsin land at $1.25 per acre. And the names on the tombstones certainly wouldn't have been those of friends of Webster to whom he, in turn, sold the holdings he had bought through a land agent in 1837.

Dr. John Porter, Webster's friend and physician, was one of the buyers. Many of his family settled what became Porter Township. Also settling there were the two Cook brothers who platted a village called Cooksville in 1842. "Waucoma" was the name the Porters chose for their adjacent village, platted in 1845. The railroad bypassed the merged hamlets, probably ensuring their later inclusion in the National Register of Historic Places.

Because there was little development after the 1840s, the place remains today an accurate reflection of villages of that era. Perhaps uniquely for Wisconsin, it has a New England flavor, with its

Cooksville Scots Pine B-Wolfgang Hoffman (1982)

grassy commons that once grazed the community's livestock. The surrounding Greek Revival style homes are constructed of lovely, vermillion-colored local brick.

Many unusual and large trees were planted early in Cooksville and were much appreciated. The old cemetery's Scots pine, at 12 feet in circumference and 50 feet in height, was for years the largest of its kind in the state. It finally succumbed to age and weather in the late 1990s. There was also a huge black locust, a mulberry, a shagbark hickory and an unusually large river birch in the town.

Sources: H. Clifton Hutchins, Madison
Larry Reed, Cooksville

The Military Road Sycamore

Fond du Lac

A combination of record size and historic location distinguish the huge sycamore growing in the city of Fond du Lac. Not only is the tree one of the largest sycamores in the state, but it has the most notable address: 100 South Military Road.

As of 2003, the giant tree had a circumference of 18 feet 4 inches and was nearly 70 feet tall. Since the sycamore is not native to Wisconsin, a specimen this large is a rarity here. Though at one time the tree faced possible removal in a road-widening project, it is now being protected by the city.

The Military Road, laid out by the United States Army from 1832 to 1835, links Fort Crawford in Prairie du Chien with Fort Winnebago in Portage and Fort Howard in Green Bay. It was the first "real" road in Wisconsin, an alternative to travel by water, and made further settlement of the state possible. The sycamore, which must be close to 200 years old, has been witness to this history.

According to Mr. Nehls, who lived in the house beneath the tree, the sycamore is a handsome presence. It is slow to leaf out in the spring, but by autumn its leaves are the biggest of any around. "When they fall, there is a big pile of them," he said with understatement.

Sources: Harold Nehls, Fond du Lac
Myron J. Medin, Jr., Fond du Lac
Ruth Worthing, Fond du Lac

Fort Atkinson's Crooked Oak

The huge misshapen bur oak on the lawn of the Congregational Church on Milwaukee Avenue in Fort Atkinson is one of a very few trees left from a once-big grove. The Fosters and the Joneses were two of the first families in Fort Atkinson and the trail linking their cabins passed the crooked oak. If travelers went off the path in the thick woods, they could always find their way back by looking for the oddly bent giant.

No one knows for sure how the tree was bent, but some say that Indian children must have played on the tree while it was still young, causing the trunk to grow in a crooked manner.

In the 1830s, the lawn under the tree was the site of the first Fourth of July celebration in Fort Atkinson. Now surrounded by buildings instead of woods, the tree stands across the street from the W. D. Hoard Company, "home of *Hoard's Dairyman,*" the outstanding dairy magazine.

When the church administrators enlarged their building in the late 1990s, they carefully protected this remarkable oak.

Source: Hannah Swart, Fort Atkinson

The Sherwood Beech

Building a road through the wilderness was one of the first and hardest tasks Wisconsin's military men faced after the end of the Blackhawk War in 1832. A beech tree, already some 200 years old in 1830 when Congress made appropriations for a survey for a military road linking Fort Crawford, Fort Winnebago and Fort Howard, had marked the ancient trail for untold numbers of Native Americans and early settlers along the east side of Lake Winnebago. In 1846 a group of soldiers resting under the tree carved their names and company numbers into the bark, evidence of their presence that the tree bore for at least another 100 years, as the village of Sherwood, in Calumet County, grew up nearby.

Soldiers of the Second Regiment stationed at Fort Howard in Green Bay, who were assigned the job of constructing the northern 56 miles of the road, undoubtedly stopped to rest under the old beech tree: their task was a formidable one.

Fort Atkinson Crooked Oak

R. D. Durbin (1981)

In most places they simply cleared a track, generally following the well-established Indian route. The southern section from Fort Crawford to Fort Winnebago was finished in 1835. Far tougher was the task of building the northern section through dense timber, and that wasn't completed until 1838. Even then it was no highway. Stumps were left in the roadbed; streams were only rudely bridged; and in boggy places, saplings laid crossways corduroyed the road. Blazed trees along the edge marked the route. About 1832, on a slight rise near Sherwood, soldiers wielding jackknives carved faces in the bark of a number of trees. Later travelers called it "False Face Hill," the name bestowed by the Indians.

On the difficult Green Bay–to–Fond du Lac section, each detachment of soldiers worked a week in turn. The 12 miles that Captain Scott and his men cleared were "as straight as an arrow and at the time was considered quite a feat," according to historian Harry Ellsworth Cole.

When finished, the road served troops and immigrants alike, Native Americans, and even wildlife who used the bridges in preference to swimming. "Green Bay races"—when folks raced each other to the land office in Green Bay to be the first to register and pay for a piece of land—provided occasional excitement. There was no stopping to rest under the Sherwood beech for them.

Famous Apple Trees

Johnny Appleseed probably never visited Wisconsin. But since apple seeds were easily carried in a small bag, many early settlers brought them to the state and planted them on their new homesteads. In 1936 the *Wisconsin Agriculturist and Farmer* asked its readers to send in nominations for the state's oldest apple tree. Fruit trees do not have a particularly long lifespan, and 60 years of age is considered quite old. Some remarkably long-lived trees were reported, however.

A man in Racine wrote that several trees planted in a small orchard on his land in 1844 were still bear-

Gays Mills: famous for its surrounding apple orchards B-Wolfgang Hoffmann (1982)

ing fruit after 92 years. A West Allis correspondent reported trees 86 years old, and others told of trees 54 years old or older.

Mark Nachtwey, from Maribel, recalled that his mother, the first white woman married in that area, planted a small orchard from apple seeds in 1842. One tree, a crab, was left and still bearing fruit in 1936, 94 years later. The tree stood at the east end of the town of DePere in Brown County, eight miles east of the city of DePere.

It was not until 1891 that the Wisconsin Horticultural Society, 20 years after its founding, first planted trial apple orchards comprising 56 varieties. Most of the early varieties are no longer grown, but two are still popular. Northwestern Greening, originated near Iola in Waupaca County in 1872, is favored for its uniform size and adaptability for freezing and canning. The Wolf River variety—described as "large, red, tender and of good flavor"—appeared in 1875 and was mentioned in the Wisconsin Horticultural Society Annual Report.

In Southwestern Wisconsin, eight or ten apple growers along the Kickapoo River, a good apple growing area, exhibited their fruit for the first time at a Wisconsin State Fair in 1905. Gays Mills is famous today as the center of more than 1,200 acres of apple trees.

Bayfield also saw early beginnings of the apple industry. Its port was accessible by boat so transportation to markets was no problem even before the railroad came in 1883. The Wealthy variety, and more recently the MacIntosh, have been widely grown on the Bayfield Peninsula, the northernmost commercial fruit growing area in the United States. In Door County, too, early horticulturists discovered the thumb extending into Green Bay had a climate remarkably suited to growing fruit, and commercial apple orchards were planted, as well as cherries and plums, in the last decade of the 19th century.

The weather along Lake Michigan from Sheboygan to Kenosha is also good for fruit trees. Some of the families in this part of the state have been growing apples for five generations.

Sources: Arthur and Olga Bassett, Baraboo

The Wyalusing Maple

The unusually large silver maple that stood on the ledge of Signal Point in Wyalusing State Park, at the confluence of the Wisconsin and Mississippi Rivers near Prairie du Chien, had a terrific view. Hundreds of feet directly below ran a bottom land segment of the Old Military Road that linked Fort Crawford at Prairie du Chien with Fort Winnebago at Portage and Fort Howard at Green Bay. The grandly bent maple was featured on early-day postcards of the park, with a man and woman stretching out their arms to span its girth of over 20 feet. Unfortunately it has long since been down. But the view is still magnificent.

The Mississippi River–to–Portage section of the Old Military Road was completed in the fall of 1835 by the Fort Crawford contingent of the Fifth Regiment of the U.S. Army, under the command of Brigadier General George Mercer Brooke. In most places it consisted of a track cleared of trees, but not stumps. While in timbered areas blazed trees marked the route, across the prairie stretches a traveler depended on a good sense of direction. Such a "poor excuse for a road" could be used only in winter when the ground was frozen or in summer when the weather was dry, according to an article by Harry Ellsworth Cole in 1925.

Wyalusing Giant Tree DNR

Source: Griff Williams, Prairie du Chien

Delavan Civil War Sign-Up Tree

R. D. Durbin (1981)

Two Civil War Sign-Up Trees

Richland and Walworth Counties

Two historic trees, one in Richland County and one in Walworth County, remind us of the grim days of the Civil War era. They served as recruiting headquarters with young men reporting there to sign up to fight for the North's cause. The Richland County tree stood alone along Highway 56 in the Town of Marshall, a short distance north of the Fancy Creek Church and Cemetery. According to old-timers, local young men became members of Company H of the 46th Wisconsin Infantry in the cool shade under the tree's branches.

The stringing of telephone wires along Highway 56 threatened the very existence of the tree, but those who loved both trees and history prevailed and saved the old landmark from destruction. Unfortunately, a strong wind blew the tree over in the late 1970s.

Another tree connected with the Civil War is a bur oak in Delavan. Located on the northwest corner of the intersection of Seventh and Washington Streets in that city, the tree is estimated to be about 250 years old and is locally referred to as the "Civil War Tree." During the years 1861 to 1865, volunteers were mustered under it before they left by rail for training camps.

Although of significant age, the Delavan Oak is not a large specimen, primarily because, for over 170 years, the surface adjacent to it has been paved. For more than a century, also, a grapevine that grew up alongside the trunk entwined itself among the branches of the tree, affording sport for young tree climbers who swung on its tendrils.

Source: W. Gordon Yadon, Delavan

Randall Wilcox Buckeye — *Green Bay Press Gazette* (1961)

The Randall Wilcox Buckeye

DePere

One of the largest Ohio buckeyes in the state towers over Wisconsin's oldest home to be continuously occupied, at 707 N. Broadway in DePere, Brown County. The Lawton family, direct descendants of Randall Wilcox, who built the home in 1836, still lives there. Growing beside the front porch, with its pillars and pediment, the tree was a spectator in its youth to the passing parade of traders and Indians, soldiers and settlers, farmers and townspeople along the Military Road from Fort Winnebago to Fort Howard.

Still in good health, the buckeye measures about 10 feet 6 inches around the trunk and is 73 feet high.

Randall Wilcox, an architect, came to the Green Bay area from the East, bringing with him all the materials he needed to build not only his home but a building across the street that became one of the first banks in the Northwest. Both buildings are in the Greek Revival style, the first appearance of this popular style in the state. It is probable that Wilcox brought the buckeye seedling with him.

As time went on, Wilcox became one of the area's most influential citizens and, in 1857, the first mayor of DePere.

Sources: Max Franz, DePere
Darlene Kronschnabel, DePere
Cathy and Randall Wilcox Lawton, DePere
Tracy Salisbury, DNR

Delavan Founder's Oak R. D. Durbin (1981)

The Delavan Founder's Oak

Another of Wisconsin's early pioneers, who kept a daily journal, recorded in it that on the night of July 2, 1836, he slept under a large bur oak tree. At the time, he was on an exploring trip in what became Walworth County. A few days later this man, Samuel Faulkner Phoenix, established a sawmill on Lake Delavan and founded the city of the same name. The next year Phoenix, a New Yorker, brought his family back to Wisconsin. He died here at the age of 42 in 1840.

The tree under which he slept is now estimated to be over 400 years old. It was a favorite roosting place for passenger pigeons from 1830 until 1870 when the birds became extinct. Its symmetrical branches extend over a circle 90 feet in diameter. Because of its huge size, spread and symmetry, it dominates the park-like grounds (next to 509 North Terrace Street), and its fellow conifers and ornamentals, with kingly majesty.

Source: Y. Gordon Yadon, Delavan

The Lone Tree Prairie Oak

St. Croix County

A sentinel standing in a sea of grasses, a large oak growing in the middle of a prairie in St. Croix County, about four miles southeast of New Richmond, served as an early-day oasis for people traveling from the New Richmond area to Hudson to buy supplies.

The tree was conveniently located about halfway along the route to Hudson, and farmers and their families would stop and rest, water their horses or oxen, and eat their noonday meal under the spreading oak's branches.

The tree's hospitality may in fact have been its undoing, for some old-timers believed that the oxen rubbing against the tree helped to kill it. At any rate, it was cut down in 1862. The well-known and historical Octagon House of Hudson, built in 1835, has a crude wooden cup made from the wood of the Lone Tree Prairie Oak.

Source: Mrs. Hugh F. Gwin, Hudson

Ripon's Signal Oak

The Signal Oak, standing at the northwest corner of Ransom and Thorne Streets in downtown Ripon, was one of Wisconsin's native bur oaks lucky enough to be growing at a hub of history. A huge tree, measuring more than 12 feet in circumference, and well over 200 years old, the oak was blown down during a tornado in the early 1990s. Along with another oak now gone, this tree served as a signal or marker tree along the famous Butte des Morts Trail from Green Bay to Oshkosh. Before soldiers from Fort Howard slashed the Military Road through the wilderness, Native Americans, fur trappers and traders, soldiers and very early settlers used the Butte des Morts trail in the winter when they could not traverse the Fox-Wisconsin waterway. Highway 44 between Ripon and Oshkosh follows part of this original route.

In the realm of social innovation, Warren Chase, in 1844, attracted devotees of the French philosopher Charles Fourier to a community called Ceresco, a short distance to the west of the tree. Some 200 followers—the Wisconsin Phalanx—built a long house there, pursuing their socialistic or communal lifestyle for seven years before disbanding.

The Signal Oak witnessed the founding of Ripon by Captain David P. Mapes in 1849, and a year later, the establishment of Ripon College on a nearby hill.

The birthplace of the Grand Old Party, a small, white frame schoolhouse within walking distance of the Signal Oak, made political history about the same time the Fourierites were disbanding. In 1852, Alvan Earle Bovay, who had been drawn to Ripon by the Fourierite communitarian experiment, had met with Horace Greeley in New York to talk about a new political party. When a bill permitting the expansion of slavery beyond the limits set in the Missouri Compromise passed the U.S. Senate in 1854, Bovay called a meeting and 53 voters gathered in the schoolhouse and organized a new party. They advocated the barring of slavery from the territories and adopted the name Bovay suggested: Republican.

Ripon Signal Oak Eric Weston (1982)

Military history, too, was played out just a short distance north of the Signal Oak when the First Wisconsin Cavalry mustered the men who would help preserve the Union during the Civil War.

Richard Dart, an early resident of the Ripon area and the son of the founder of Dartford (Green Lake), researched the historical facts about the Signal Oak. Lyle Cors wrote a tribute to the tree in 1956, and the first in a series of Ripon historical prints, an idea conceived by the Community Art Center, commemorated the city's oldest landmark with a handsome, limited-edition silkscreen print.

Sources: Esther Bent, Ripon
Doug Lyke, Ripon
Allan Mortenson, Ripon
Donald Redman, Ripon
Tracy Salisbury, DNR

The Pleasant Company Oak

Middleton

There is something about an ancient oak standing alone in a grassy field that speaks to strength, perseverance, creative inspiration, generosity and grace. It was around such an oak in Middleton, Wisconsin, that Pleasant T. Rowland chose to build the corporate headquarters of Pleasant Company. Pleasant, a former educator, had an idea for creating products that would foster pride in the rich traditions of growing up female in America. Between the time that she founded the company in 1986 and sold it to Mattel, Inc., the world's leading toy maker, in 1998, Pleasant Company sold more than 61 million books and 5 million American Girl dolls. They were received by a generation of young girls, whose lives were forever enriched. The Pleasant Company Oak, just outside Pleasant's office window, was an anchor to this enterprise, witnessing remarkable business growth and providing inspiration to the creative staff. Its silhouette was seen on company stationery and marketing literature. During the many phases of building expansion on the site Pleasant hired Allison Tree Care, Inc., to care for the tree, making it quite clear that it was an integral part of the company and she would tolerate no harm to it. The tree remained healthy and continued to grow from year to year along with Pleasant's dream. Pleasant used the oak as a symbol of the company's mission in the following words addressed to her employees:

In the middle of the field in front of Pleasant Company stands an enormous, near-perfect oak tree with beautiful, leafy branches reaching to the sky. That magnificent tree began from a single seed long, long ago. I am glad that tree is in our front yard, for as time-worn as the metaphor may be, it is a reminder of the constant truth that "mighty oaks from tiny acorns grow," that dreams planted deep and nourished well will flourish in goodness for years to come.

Pleasant Company Oak Photo courtesy of American Girl

So, too, with Pleasant Company. What was a tiny seed of an idea in my mind years ago, an acorn of inspiration, has grown into all that surrounds us today. What we have built together is not yet a mighty oak, but it is certainly a sturdy sapling surging with energy and promise. With our continued hard work and unyielding commitment to quality, Pleasant Company will grow to be a mighty force—an example to the toy industry, perhaps to many American businesses, that serving the needs of children is a more powerful mission than serving the bottom line.

It was under the oak that Pleasant gathered her entire staff to announce the transfer of the company to Mattel and to award remarkably generous bonuses to her employees. That gracious sense of giving and generosity continued with profits from the Pleasant Company sale contributed to the Madison Children's Museum, the Overture Center in downtown Madison, the Great Performance Fund for the Arts and many other organizations devoted to the advancement of the human spirit. The Pleasant Company Oak stands today as solid and graceful as ever, an inspiration to all who see it as a symbol of the company's accomplishments and a tribute to the power of dreams—the tiny acorns from which mighty oaks grow.

Chapter III

Hanging Trees

Durand Lynch Tree Melba Baehr (1962)

Durand Lynch Tree

Pepin County

An old oak tree on the lawn of the Durand Courthouse in Pepin County figured in the lynching of Ed Williams back on November 19, 1881. Ed and his brother Lon, who specialized in stealing horses but dabbled in other crimes—such as bank robbery and murder—were sometimes compared to the James Brothers. The story was written up by Christine Klatt, an early Dunn County settler, and was published by the Dunn County Historical Society.

When two lawmen, ex-sheriff Charles Coleman of Pepin County and his brother Milton, who was under-sheriff of neighboring Dunn County, tried to arrest the two Williams brothers in Durand, the sheriffs were both shot dead on the spot before witnesses. The subsequent chase of Lon and Ed Williams ended in Nebraska with the capture of Ed. Lon escaped.

They brought Ed Williams back and jailed him in Menomonie, then took him to Durand for preliminary examination. When they debarked from the ferry at Durand, they found about 50 people waiting. The crowd muttered a few threatening remarks but no real trouble ensued.

At two o'clock the next afternoon, Ed Williams appeared in court. Men, women and children from nearly all parts of Pepin and adjoining counties crowded the room, eager to get a view of the "notorious criminal." He gave his name as William E. Maxwell (Ed Williams was an alias), pleaded not guilty to the charges, waived an examination and was committed for trial. The officers started to lead him back to his cell. Then someone in the crowd yelled, "Hang the son of a bitch," and "a dozen or more determined men tackled the officers."

"The officers made a desperate resistance," and "Maxwell fought like a tiger," but to no avail. A noose with a hangman's knot was slipped over his head, and he was dragged out on the porch and across the lawn to the oak tree "and quickly suspended in the air, with his handcuffs still on and a heavy pair of shackles hanging from his left foot."

After about 15 minutes, the officers were able to get to the body. They cut it down, and it was later interred in the Potter's Field of the Durand Cemetery.

The community did a bit of soul-searching after the episode, debating the issue back and forth in the local newspapers. An item in the Eau Galle paper noted that "A more brutal affair never happened anywhere and ... it is a reflection upon the morality and civilization of Northwestern Wisconsin."

The lynching continued to be debated in the press as Durand took on a reputation as a "Hanging Town." It was also accused of not having borne its share of supplying men or provisions during the hunt for the Williams brothers. The Menomonie press smugly noted that while Ed Williams was in the Menomonie Courthouse, surrounded by a dense throng of people, nearly every one of whom was a personal friend of the murdered Coleman brothers, and with no guard but the two or three officers in charge, he "was yet unmolested and so far as personal violence was concerned, was in perfect safety. "

The *Durand Courier* attempted to put the case in perspective. "Finally, we think we can truthfully assert that Ed Maxwell was hung, not in the spirit of revenge, for the crimes he had committed, not to show the bravery of the community ... but on future protection to this place and other communities. A man who not only boasts of his past misdeeds but expresses determination to continue on his evil course, and asserts that he will not allow any number of lives to stand between him and his liberty, is too dangerous a citizen to let live, and any man or number of men who wipe such a dangerous desperado from the face of the earth should be commended instead of condemned."

The *Knapp News* added a last word: "There is seldom an evil done but there can be some good traced from it. So in the hanging of Ed Williams, we believe that the effect will be salutary on all those who might have been ambitious of following in his footsteps. The boys who entertain a desire to become bank robbers will now hesitate lest they meet a similar fate."

The hanging tree was cut down in 1977. It was dying and was considered a hazard to pedestrians.

Sources: Melba Baehr, Eau Claire
Christine Granger Klatt, Menomonie
Howard Mead, Madison

The Parade Day Hanging Tree

La Crosse

On October 16, 1884, Frank Burton, a prominent La Crosse citizen, was marching at the head of a political parade in that city. Along both sides of the parade route hundreds of onlookers stood and watched. Suddenly, as the parade passed the intersection of Fourth and Main Streets, a man in the crowd, Scotty Mitchell, raised a gun, fired and shot Burton down.

The assassin was immediately seized, taken into custody and placed in the county jail a block away at Fourth and State Streets. Within hours, a mob began to form, became increasingly unruly and forced their way into the jail. They dragged the assassin Mitchell outside and hung him from an oak tree on the south side of the Court House.

The tree, of course, was promptly dubbed the Hanging Tree, but the appellation was short-lived because the tree was ordered cut down almost at once. The city officials did not want the public to remember that mob rule had taken precedence over due legal process. Old-timers who heard the story from their elders, however, still describe the incident.

Source: Myer Katz, La Crosse

The Accidental Hanging Tree

Trempealeau County

An oak tree in Trempealeau County figured in a not-so-warm welcome home for a man just released from prison. In fact, so unwelcome was he, his family and neighbors gathered to scare him away for good. While his wife served coffee to guests, they pretended to hang him. But things went awry and the scare turned into a real hanging for Hans Jakob Olson, aged 51, on the night of November 24, 1889. The oak tree that served as gallows stood on the John Brekke farm in the Town of Preston. It has long since been taken down.

The rascally Olson, who was greatly feared by those who knew him best, had the reputation of marauding around the community at night, scaring people half to death. He had just returned to his farm after serving a term at the state prison at Waupun for attempted arson and murder of a local hardware dealer to whom he owed money. (He had bored holes in chunks of stove wood, filled them with gunpowder and had them delivered to the hardware dealer. The stove exploded but no one was hurt.)

The hanging, while intended only to convince Olson to leave the area, "accidentally" ended his depredations for good. The victim's grown son and two neighbors and Olson's wife, Bertha, were directly accused of the crime. Some 30 other men were implicated as spectators who did nothing to stop it. The four convicted received sentences of life imprisonment but were pardoned after five years. The spectators were each fined $100 plus $7 costs.

Source: Odin Anderson, Madison

The Souvenir Hanging Tree

Vernon County

Andrew Grandstaff, who murdered four people, two of them children, was hanged from a tree on the Vernon County Courthouse lawn before he could be brought to trial. It happened on the night of Viroqua High School's graduation—June 28, 1888.

Afterwards the tree, virtually destroyed by souvenir hunters, had to be taken down. A portion of it, as well as the victim's clothing, is still on display at the Vernon County Historical Society in Viroqua.

A lifelong Kickapoogian, Grandstaff had worked for numerous people in the area, including Mr. and Mrs. Reuben Drake. When he heard that the Drakes had come into some money, he broke into their home and killed them and their two grandchildren. Discovered, he ran away on foot but was caught near Liberty, between Viola and Viroqua. Apparently he hadn't found any money, but he claimed to have an accomplice—who never turned up.

So incensed was the community by the crime that a mob broke into the jail and seized Grandstaff. People hearing about the affair came quickly from all over the area to witness the summary execution. The ringleaders of the lynching asked Grandstaff if he had a last request. He replied, "A chaw of terbaccer." Reportedly it was granted before they strung him up.

The Drakes and their grandchildren lie in the Sugar Grove Cemetery on Highway 14 between Readstown and Bosstown. As for the murderer, a very old-timer remembered that as a small boy, he stood on his front porch and watched as the wagon carrying Grandstaff's body in a wood coffin went by his house, en route to the paupers' section of the old County Home burial plot. There are, however, no official records of the interment.

Sources: Carol A. Oliver, Viola
Connie Stevens, Viroqua
Mrs. Phil Tate, Viroqua
Rhoda White, Viroqua

The Janesville Hanging Tree

The kind of tree was not recorded, but it served as a hanging tree for David Mayberry, a convicted murderer with a record of criminal offenses. His victim was Andrew Alger, a Jefferson County man who was working as a logger in Beloit. In Mayberry's presence, Alger made the mistake of bragging about how much money he had made from selling the timber he cut.

Mayberry immediately joined up with him and they traveled to Janesville, where, unknown to Alger, Mayberry purchased a hatchet. They continued on toward Milton and, just three miles short of the town, Mayberry found his opportunity and hacked Alger to death.

Arrested in Rockford, Mayberry was brought back to Janesville for trial. His subsequent conviction in July 1859 stirred a lynch mob to action and they broke into the jail. They were successfully opposed by another faction who felt the course of justice should not be interfered with. The next day, however, the hanging mob returned and this time managed to drag Mayberry from his cell. They hung him from a branch of the tree that stood in the open space in front of the courthouse.

Later, someone cut the tree down, perhaps because of its disconcerting effect on the prisoners.

Source: Maurice Montgomery, Janesville

Chapter IV

Homeplace Trees

Jones Maple

Edward C. Jones (1982)

The Jones Maples

Fort Atkinson

An old maple, planted on the Jones Dairy Farm in Fort Atkinson, has great meaning for the Jones family, one of the earliest to settle in the area.

The huge tree is called the Centennial Maple. Icabod Cole, of Michigan, who was at that time visiting his daughter Mary Frances—wife of Milo Jones—planted it in 1876 in commemoration of the nation's 100th birthday.

Exactly 100 years later, Edward C. Jones, grandson of Milo and Mary Frances Cole Jones, planted two Bicentennial maples, both seedlings from the Centennial Maple, to carry on the family tradition.

The first Milo Jones worked as a surveyor for the federal government from 1832 to 1836 in what was then Wisconsin Territory. Surveyors had their choice of money or 1,000 acres of free land in payment for their services. Jones took the land. The sixth generation of Joneses is living on it today and the farm is listed in the National Register of Historic Places.

Speaking of the Centennial Maple, Mrs. Edward C Jones wrote: "It is by now a beautiful large tree that has been well cared for during the years. Under its shade we often have family picnics in the summertime."

The Jones Dairy Farm has long been well known in the state. The Jones sausage business was started by Milo Jones II in the 1880s. Its products are now available worldwide.

The dairy farm, the sausage business, the beautiful, well-cared-for grounds with handsome residences and lovely old trees reflect the owners' rightful pride and faith in family, tradition and their state and country.

Sources: Mr. and Mrs. Edward C. Jones, Fort Atkinson
Hannah Swart, Fort Atkinson

The Magnus Swenson Trees

Madison

When 14-year-old Magnus Swenson came to America, barely surviving the sea voyage in a rudderless ship, he brought with him seeds from a forest of beech trees near his home in Norway. And he eventually planted them on land he bought bordering Lake Mendota, then outside the city of Madison, which he called "Thorstrand" or "Tor's Beach." Along with the beech trees, he also planted elms, Norway spruce, and balsam fir. The beech trees grew to maturity, forming a lovely grove, but the climate was not ideal for them and they died some years ago. The property also included a very old elm, spruce, and fir as well as 30 sassafras trees, which are considered a rarity in this northern climate.

Immediately after his arrival here, Swenson went to Janesville where he worked for an uncle repairing train locomotives. Later, he took degrees in both engineering and chemistry at the University of Wisconsin–Madison, the first to combine those majors. As an engineer he built the Kilbourn dam in Wisconsin Dells in the first decade of the twentieth century, as well as a dam at Prairie du Chien. He also became an inventor of note. Among his concerns, way back then, were ways factories could prevent pollution and use byproducts. Swenson was a friend of the Norwegian author Ole Rolvaag and active in Madison Norwegian organizations.

His original farm has been cut in two by University Avenue, but Thorstrand and Magnus Swenson Drives still indicate its location.

Source: Polly O'Hare, Madison

Magnus Swenson Norway Spruce

B-Wolfgang Hoffmann (1982)

Lodi Oak B-Wolfgang Hoffmann (1982)

The Lodi Oak

There's a huge, picturesque old white oak on the former Carberry place on Millston Avenue in Lodi which, its owner firmly believes, should have won an essay contest as "My Favorite Tree." The Lodi Parks Commission, in cooperation with the Lodi *Enterprise,* sponsored the 1977 contest to inspire people to plant more trees. In the process, the *Enterprise* won an award for public service offered annually by the Wisconsin Association of Manufacturers.

The Carberry tree came in third in the contest. It is a tree that "everybody in Lodi knows," and many generations of children have climbed up into its capacious branches and built tree houses there. A boy with a new ax once tried to cut the tree down. Luckily, Mrs. Carberry was home and heard him. The scar healed.

The property on which the oak grows was purchased 50 years ago by Mr. and Mrs. Carberry and is still in the family. They bought it because they liked the tree, estimated to be about 300 years old. Their son, as a child, used to climb the oak. Mrs. Carberry herself climbed the tree into middle age.

William Sparks, who started the phone company in Lodi, once owned the tree, and his daughter Catharine remembered playing and eating sandwiches under it. When a tree becomes so firmly a part of people's nostalgic memories, it doesn't have to win first prize in essay contests. As Mrs. Carberry concluded her contest entry: "May it live on, and when we are gone, may others enjoy it as much as we, and as much as scores of people have in the past."

Sources: Mrs. Robert Carberry, Lodi
David Skoloda, Lodi
Robert Guethlein, grandson of Mrs. Carberry, Lodi

The Poet's Larch

Iowa County

Edna Meudt, one of Wisconsin's premiere poets during her lifetime, had a deep and intuitive appreciation for the trees on her Dodge County farm. When a 1973 Saturday tornado uprooted a large old European larch and blew it completely across the road, she didn't just consign the tree to a firewood fate. She was determined to save the century-old specimen, knowing that it ranked among Wisconsin's champion trees.

But more than that, she remembered that in March 1935, when her family had moved to the farm, she thought the three larches on the property were dead pines because they had cones, but no needles. She decided in early May to remove them. Then, she writes, "That first week I noticed a kind of green mist on those trees. Two days later there was the airiest green cast and then the fine needles. How gorgeous they were to become a week or so later! The new cones are fuchsia, about the size of big pussywillows. Unbelievable! The next Sunday I took a sizeable bouquet to my father (Grandfather was a forester in the Sumava in Bohemia). He identified them."

After the 1973 tornado had wreaked its havoc and departed, she called her friend, Reid Gilbert, who at that time ran the Valley Studio near Spring Green, for help. On Sunday, the well-known actor and mime brought 26 apprentices to the Meudt farm to help clean up the debris of downed trees left by the wind. When Mrs. Meudt indicated she wanted to save the larch, some of the apprentices expressed amusement. She told them about the spring and summer evenings, before farm ponds and mosquitoes, spent in a hammock under the tree. She told them about

Poet's Larch B-Wolfgang Hoffmann (1982)

the good talk and courting that had gone on beneath its limbs. One of the apprentices, Cleasius Goldman, who said he was half Crow Indian, warned that they must let the tree spirit know that help was coming or it would leave the tree. At his request, Mrs. Meudt supplied him with a bedsheet, which he ripped into strips and tied to the roots of the tree. Then they covered the roots with a blanket and watered it well.

On Monday the crane operator arrived with his rig to replant the 80-foot tree as "the strips tied to the tree's roots waved crazily in a rising wind." Sudden new tornado warnings made it necessary for him to return to town. He came again the next day, however, and lifted the tree back into its place, bracing it with steel cables and supports. Although county forester James Widder estimated the tree's chances of survival at less than 40 percent, the larch did survive. The tree spirit had not departed, for the larch put out new needles next year in its familiar, breathtaking spring display of beauty.

But its ordeal was not over. In July 1974, Mrs. Meudt was standing on her porch during a rainstorm when she saw a huge fireball drop from the sky, sit for an instant on the tip of the larch, then roll down the side and disappear into the ground, along the path of the steel support. It peeled off the bark all the way down, but the remaining "half a tree" continued to thrive.

Whether the tree's second chance at life was the result of its vibrant spirit or that of its poet owner, is hard to say, but it still exists. As an illustration of what can be accomplished by people who care about trees, it has a fitting poetic mysticism.

Source: Edna Meudt, Dodgeville

Meier's Oak

Price County

Some trees may not be historic in the most limiting sense of the word, yet their stories throw small but illuminating shafts of light on the times in which they lived and the people who "owned" them. Such a tree is the Big Oak that may still stand on the Meier homestead in Price County, near Ogema.

It was estimated to be over 300 years old, 100 feet tall, with approximately 1,330 board feet of prime lumber. Here is its story, told mostly in the words of Roy Meier, whose father first homesteaded the land in 1883.

> *Around 1860 a tote road starting in Jenny (Merrill) went north to haul supplies to the pine camps at the headwaters of the Spirit, Jump, Chippewa and Flambeau Rivers. This oak is located where the tote road crossed the Spirit River.*

Timber cruisers sometimes called it the Squirrel River, because the grove of white oak harbored so many squirrels. Locally the trees were called blue oak because when the steel of the ax or saw came in contact with the wood, the steel turned blue.

Originally, the oaks had no value as lumber because they could not be floated to market. Around 1900, Roy Meier's father cut a tree and had it sawed into planks to be used in making sleigh runners. In 1907, he used and sold many of the rock elm, which were mixed in with the oak, for barn timbers.

> *About 1910, the M. T. & W. Railroad came as far as our farm and a shipbuilding company from Manitowoc came looking for oak timber. My father offered to sell the oak trees, except for the one largest tree. He admired that tree and wanted it to stand and send out acorns.*

Meier's Oak — Roy Meier

As it turned out, the deal fell through anyway, because the company was unable to find more oak in the area. After Roy's father died in 1919, Roy was especially glad the large old oak his father had loved still stood.

By 1924 "Roddis of Marshfield started buying veneer in the area," so Roy started cutting rock elm and some of the oak. "But of course, we would not cut the big tree."

Then came the Depression and hard times, and no one was buying trees. By 1940, however, the veneer companies were again looking for good timber.

> *We were very short of money when Peterson of Park Falls came looking for veneer logs. He offered me $100 for the Big Oak. I went to the house and asked my wife. She said, "We need so many things, but let's try and get along without selling the Big Oak."*
>
> *There are perhaps a dozen small oak trees coming. We think they were started from acorns from our Big Oak. Lightning has hit it once but only made a streak down the side. This has healed over. So it will keep producing acorns, and the squirrels will bury some for seed.*

Source: Roy R. Meier, Ogema

The Rufus Dodge Maples

Fort Atkinson

The line of maple trees just at the edge of McCoy Park in Fort Atkinson was planted by Rufus Dodge, one of the early settlers. Dodge and his family immigrated from New York in 1837 and started farming about a mile west of the city on Highway 12. At that time the highway was known as the Territorial Road and connected Fort Crawford in Prairie du Chien with Mineral Point and Milwaukee.

Dodge married Rebecca Foster, sister of Dwight Foster, the city's first settler, and indulged his love of trees by planting a row of maple seedlings along his road. People are still enjoying their beauty and shade.

One of the trees is irregular in shape because of a shooting accident. A relative of Dodge's shot at some blackbirds sitting in the tree and tore away a number of branches. The gap is still noticeable.

Also on the Dodge property is a pear tree, planted over 100 years ago, that is still bearing fruit.

Sources: Hannah Swart, Fort Atkinson
Tom Kramp, Fort Atkinson

Rufus Dodge Maples R. D. Durbin (1981)

The Grant County Sycamore

The farm is named Sycamore Crest Farm and the road is called Sycamore Road. And, yes, there is quite a story to the old sycamore.

One of Grant County's earliest settlers, Joseph Crain Orr, went south after the Civil War to find his son, who had been taken prisoner. Orr found him, but unfortunately, the young man was near death from his wounds and the effects of his long imprisonment. Though moved to a hospital in Tennessee, he died.

Orr, accompanying his son's coffin, returned to Wisconsin via his original home state of Ohio. While there, he chose a sycamore seedling to bring back to the new homestead. He planted the young tree in the yard, just south across a field from the local cemetery where he buried his son.

The massive sycamore was "discovered" shortly before its 100th birthday by the seventh and eighth grade classes at the nearby North Andover School near Bloomington in Grant County. They measured the tree and wrote to the State Conservation Department (now the Department of Natural Resources) suggesting the tree's inclusion among the state's record trees. Shortly afterwards, the tree gained official designation as the largest sycamore in Wisconsin at that time.

Source: Craig Tiedemann, Madison

Grant County Sycamore B-Wolfgang Hoffmann (1979)

The Damuth Civil War Maples

Jefferson County

The Damuth maples, growing along Curtis Mill Road just across from the Rock River on Highway 106 east of Fort Atkinson, were planted during the Civil War while the young men in the family were serving in the union army.

George Damuth started the farm; later his sons Sam and Dolph took over the home farm and another farm adjoining it. When the Civil War broke out, Sam and Dolph went off to fight for the Union. While they were gone, the women and children did their part to keep things going. Eighteen-year-old Martha Damuth and her nephew gathered young hard maples from the woods and set them out in a long row bordering the road between the two farms.

They must have been a welcome sight for Sam and Dolph when they finally returned from battle.

Source: Hannah Swart, Fort Atkinson

The Rhodes Bald Cypress

Kenosha County

A bald cypress, a tree normally found in the semi-tropics, has achieved record size in Wisconsin because of a man's penchant for the unusual. Well over 140 years old, it still stands in the yard of the family farm in Kenosha County. Phil Sander, a conservationist and naturalist, rediscovered the tree and ferreted out the story.

Clarence Rhodes, who with his brother Frank owned the farm until it was taken over by the Bong Air Base, said his grandfather, John Rhodes, planted the tree near the house, along with at least 14 other unusual species of trees, because he liked unusual things. The tree produces round cones instead of oblong ones.

John Rhodes, whose father, Abraham, had bought the land for $1.25 an acre back in the 1840s after sailing to the United States from Liverpool, England, obtained the tree when he visited his daughter, Mary Hoyt, in Estherwood, Louisiana. Possibly the tree symbolized the healing over of the rift that developed when his daughter married young Hoyt, contrary to her parents' wishes.

The tree, which now stands over 65 feet tall and measures 11 feet 5 inches in circumference, is the largest of its species in the state. Typically, strange nubbins, called knees, shoot up from the roots of cypress trees. It is recorded by John's son, Clarence, that the knees were a nuisance and had to be continually cut away so the lawn could be mowed.

The tree revealed its preference for warmer climes by audibly suffering through the cold Wisconsin winters. The Rhodes family, inside the farmhouse on nights of bitter cold, would often hear the tree give a mighty crack. In the morning, they would find a crevice in the trunk big enough to put a mittened hand in. The cracks healed over, though, forming protrusions on the trunk, and the tree lived on.

Cypress trees are known for their ability to attain a singular age, and for their importance in history and in the Bible. Perhaps those were the associations that appealed to John Rhodes when he brought the tree back from the south and planted it on his farm. Four generations lived there. The bald cypress and its unusual fellow trees, including a white ash, an Austrian pine, and a ginkgo, are now part of what is called the Richard Bong State Recreational Area,

Rhodes Bald Cypress with Phil Sander Wally E. Schulz (1962)

named after Major Richard I. Bong, a Wisconsin native who was America's leading air ace during World War II.

Source: Phil Sander, Kenosha

The Walking Staff Tree

Sauk County

Back in 1848, it might take a settler, lured by promises of cheap, fertile land, two years to travel from the East via covered wagon. The Jacob Tinkham family from Vermont is one that made the arduous trek. Along the way Tinkham cut a walking staff from a cottonwood tree to ease the miles of foot travel. When he got to his destination—land in Sauk County between Lake Delton and Reedsburg along what is now Highway 23—he drove the staff into the ground. It grew into a superb tree, some 23 feet in circumference.

A subsequent owner of the land donated the acre on which the tree stood for a roadside park, named in honor of the cottonwood. But the tree attracted lightning as well as visitors, and eventually road officials felt the hollow tree should come down for safety's sake. A resident of the area, the late Hilbert Kleeber, and the Reedsburg Boy Scouts planted a replacement cottonwood. Unfortunately, the state has since closed the park, and the land reverted back to the owner.

Mr. Vernon Harrison, the town clerk who had negotiated the park plan with the state in the 1950s, saved a portion of the old cottonwood. He presented it to the Reedsburg Historical Society, along with a letter from Jacob Tinkham's grandson describing the origin of the tree. A shoot from the roots of that original tree is now planted in Mr. Harrison's front yard, a still living symbol of one pioneer family's covered-wagon journey to Wisconsin.

Source: Vernon Harrison, Reedsburg

Kohler Boxelder at Riverbend with Mrs. Walter Kohler, Sr.

Julilly Kohler Hausman (c. 1929)

The Kohler Boxelder

The Kohler Company, of Kohler, has a soft spot for trees. Along with numerous other philanthropic pursuits, the Kohler family has long been engaged in an active program of tree planting, in the town, on the grounds around the company buildings, and on their various personal estates.

In the 1920s, an old and unusually pendulous boxelder tree served as the focal point around which former state governor Walter Kohler, Sr., built his home, "Riverbend." A huge, spreading tree, with branches dipping down to the ground and rooting themselves, the tree shaded an area more than 108 feet across. Unfortunately the tree began to die in the late 1970s and is now gone.

Sources: John Green, Kohler
Bill Reichenbach, Kohler

The Hoard Museum Pine

Fort Atkinson

The Austrian pine that grows beside the front porch of the Hoard Historical Museum in Fort Atkinson was planted about the same time Rufus White, a Fort Atkinson druggist, built the house in 1864.

The second owner of the home was Frank Hoard, a member of the illustrious Hoard family that included William Dempster Hoard, who was elected governor of the state in 1888 after founding *Hoard's Dairyman* and encouraging farmers to "substitute the cow for the plow."

The Hoard Historical Museum features one of Wisconsin's outstanding collections of local history.

Source: Hannah Swart, Fort Atkinson

Hoard Museum Pine R. D. Durbin (1981)

Trees That Stood in the Way

According to Walter E. Scott, noted discoverer and cataloger of Wisconsin trees, "Mature men wept to see the desecration of beautiful trees during the 1950 road-widening project on Highway 30 in Waukesha County."

Mrs. Otton L. Prime, a woman who lived along that road, did everything she could to save what was described as "the most beautiful hackberry" in the state and probably the largest, as well as oaks, maples, black walnuts and butternuts. All were over 100 years old, some 200 or 300. Angered spectators, brought together by the incident, decried the callousness and undue haste with which the destruction of "that scenic, mile-long arborway" was destroyed.

Road-widening projects are still taking their toll of old trees, and often whether cutting is necessary or not is a matter of personal opinion. Some years ago, Mrs. Herbert Mohns of Brodhead offered to pay any price for the parcel of land on which a 300-year-old roadside elm, 21 feet in circumference and over 100 feet high, stood. The owner refused her offer with a rude "The g—d— tree has to come down!" and proceeded to cut it. Mrs. Mohns later wrote an "Ode to the Old Elm Tree" that was published in *Prairie Farmer.*

The Monona Drive widening project in Madison caused the destruction of several trees, among them an old bur oak, despite the objections of residents led by Mrs. F. I. Elliott. In a plea before the city council, she stressed the aesthetic value of trees to the community.

Around 1980, several old prairie oak trees along Wilkinson Road in Dunlap Hollow in the Township of Mazomanie in Dane County fell to ax and saw. These giants predated the arrival of the settlers who found their way into the groves of open-grown oaks and built their homes of native limestone near their shade. The cutting included a huge black oak whose stump was almost 13 feet in circumference. An added legal question involved in this case was who gets the wood when trees are cut along roads on private property?

Folks sometimes go to great expense to preserve a tree they love. Many years ago, a man named Harry Netherwood lived at the corner of Garfield and Chandler Streets in Madison. He liked to look out the windows of his home at the beautiful, large bur oak growing in a vacant lot across Chandler. When the lot went up for sale in 1911 he bought it, and subsequently paid taxes on it every year until 1942. Netherwood, who had enjoyed the tree's presence for 50 years, undoubtedly thought the cost well worth it. Eventually as an old man he had to sell his property, and the tree was, of course, cut down to make way for a new building.

In 2003, Eileen Potts Dawson, as a Friend of the Madison School Forest in Verona Township, Dane County, fought valiantly to save ancient twin oaks from a road-widening project. The battle lost, she memorialized the trees in the poem *In Memoriam, July 23, 2003,* excerpted below. It speaks for all those who lament the loss of trees that stood in the way of progress.

They survived the footfalls and the hooves,
the sharp-clawed paws, the insects and the droughts,
the lightning, the winds, the wagon wheels.
They spread their yearling limbs while Henry David Thoreau and John Muir were writing
of the splendor, the fragility of nature, and our responsibilities for caretaking.
The oaks grew tall and strong and straight,
while near and far other woods disappeared, farm fields sprouted rows of green and gold
Concrete described a little roadway.

They survived, they thrived, they flourished.

In time, standing side by side, these twin oaks became
Sentinel trees for all who came to walk their forest home.
They were the symbols of what lay just ahead, down one path or up another,
this forest of black oak and white, bur oak and red, hickory and cherry, sugar maple and pine.
Survivors
asking nothing of us except respect for their being, giving us so much in return.

Sources: Mrs. Dennis Lenzendorf, Mazomanie
Eileen Potts Dawson, Madison

Chapter V

Indian Trees

Indian Half-way Tree — Rich Rygh, *The Capital Times* (1980)

The Indian Half-Way Tree

Green County

To the Indians, the large bur oak on the farm located on old Highway 81 just past the Brodhead airport in Green County marked a point halfway between Lake Michigan and the Mississippi River. In the 1800s, every spring, according to the old-timers, the Indians, possibly Oneidas, passed north along the Sugar River. They usually divided their party, some ascending the river in canoes, others riding horseback or walking along the bank. The Half-Way Tree naturally was a favorite camping spot for the Native Americans and became one for early white settlers as well.

Charles M. Warner, who lived on the farm then, remembered that one year the Indians had with them a 100-year-old woman. She had a tent all to herself in deference to her age. The Indians made their last trip in the spring of 1878, and there were only two of them, a father and son.

The Warner family acquired the 200-acre farm in 1857. About ten years later, young Charles saw an Indian chief standing at the door of his father's blacksmith shop. Pointing to the oak tree, the chief made them understand that it should never be cut.

It never has been. The tree has weathered some vicissitudes, however. Its bark has been much hacked by tomahawks and one year a drought dried up nearly all the leaves.

A marker placed by the road points to the tree 300 yards away. It reads: "This bur oak marks the halfway point between Lake Michigan and the Mississippi River, paced off by Indian runners and confirmed by a U.S. survey in 1832."

According to present-day experts, the tree is off by about six miles, but there is disagreement about the methods of measurement. At any rate, the old tree still stands, protected through the years by various owners, and still marks the half-way point, or close enough to it, along the ancient Indian trail.

Sources: Lillian C. Allen, Brodhead
Dorothy Gessert, Brodhead
Michael J. Williams, Milton

The Cameron Council Oak

Oshkosh

Living with a big tree makes it easy to give directions. The Stevensons, in Oshkosh, used to tell people their house was next to the big bur oak because everyone knew where that was. Already mighty before the early settlers came to Wisconsin Territory, the Cameron Oak was a favorite council tree for the Native Americans in the area.

Known as the Cameron Oak for a former owner of the land, the venerable tree was estimated to be over 350 years old. It was originally part of an oak grove cut down to make room for the campus of the University of Wisconsin–Oshkosh.

Bars bolted through the two main trunks and three support wires helped stabilize the monarch. Even though it was regularly trimmed of dead wood and fertilized to keep it in top condition, it came down during a storm in the mid 1990s.

A false story that a man was murdered under the tree when it was still part of the original oak grove circulated for years. However, the great-granddaughter of the murdered man recently set the story straight. Her ancestor was not killed under the oak but some distance away at the site of what is now Blackhawk Commons.

Deprived of connection with a grisly event, the tree was dramatic, nevertheless, in its beauty. Birds' nests hidden among the branches brought new life to it each spring; squirrels lived there all year long. In the fall the tree produced bushels of leaves and barrels of acorns. It was a favorite stopping point for school children on field trips. The urge to climb it was sometimes irresistible and occasionally the owners would look up to see college students out on limbs.

D. N. Cameron remembered climbing the old oak as a boy to watch a circus performance in a nearby clearing. He was responsible for saving the tree during the ice storm of 1929, setting up a dozen wooden supports to prevent the ice-laden branches from breaking.

Sources: Allen L. Singstock, Oshkosh
Sandra Taubel, Oshkosh
Rod Brusius, Oshkosh
Tracy Salisbury, Green Bay

The Indian Game Trees

Winnebago County

Coffee trees in Wisconsin? A grove of Kentucky coffee trees on the farm of Minor Harris on the north shore of Rush Lake, Nepsuskun Township, in Winnebago County began growing over 100 years ago. How did these trees, which are native to the southern states, find their way to Wisconsin?

Indians. It would appear that the Native Americans who passed through the area brought coffee tree pods with them, and the result is a natural Kentucky coffee tree plantation containing nearly 50 individuals.

Why were the Indians carrying around coffee tree pods? Because they used them as dice in a gambling game. The seeds probably came from players who camped on the banks of Rush Lake. The seed pods are six to eight inches long and contain large seeds about the size of shelled hazelnuts, only flatter.

The man on whose land the coffee trees are growing never knew what they were until two horticulturists told him. He had been wondering for over 50 years. There are larger individual specimens of Kentucky coffee trees growing in Wisconsin. But the Harris farm boasts the only plantation in the entire state.

Blackhawk Country Club's Spirit Oak

Madison

Members and visitors alike loved the Spirit Oak on the grounds of the Blackhawk Country Club in Shorewood Hills, Madison, because of its stately beauty as well as the history it reminded them of. It did not live, as they hoped it would, to celebrate the Bicentennial. In September 1974 the tree fell because the roots were no longer able to support the massive trunk, 12 feet 2 inches in circumference, and heavy branches. By counting the rings in a cross section, dendrologists established its age at 227 years.

In its youth the Spirit Oak lived with Indians passing along the southwest shore of nearby Lake Mendota. They left a rich heritage, and the Spirit Oak's neighborhood is marked with trails and campsites used by the Sauk and Fox tribes. When the tree was about half a century old, Sauk warrior Black Hawk signed the treaty to give up his homelands around the Rock River and agreed to move to territory west of the Mississippi. He left in 1831 but returned the next year, dissatisfied with the new lands and determined to reclaim his Rock River homelands. As he said at an 1831 Council: "I wish to remain where the bones of my fathers are laid."

But anticipated support from other Indian tribes did not materialize, and he was forced to retreat, passing near the Spirit Oak along the west shore of Lake Mendota as he headed toward Wisconsin Heights. In the bloody battles that followed there and at Bad Axe on the Mississippi, most of his followers were either drowned or massacred by the whites or their Indian mercenaries.

Disaster that it was, the defeat of Black Hawk served to open the area to settlement. First to come were the surveyors, followed by the homesteaders, and the white people's ways rapidly began to supersede the Indian culture.

The railroad came in 1867 when the Spirit Oak was 120 years old, passing near the middle-aged tree. When it was a venerable 175, the Blackhawk Country Club was founded, and a new era of care and preservation made the tree's last decades a period of increased growth.

Although the Spirit Oak is no longer standing, two cross-sections were preserved to be saved among the treasures of the club. Honorary club president L. J. Markwardt also lovingly prepared an "autobiography" of the tree.

The Blackhawk Hickory

Madison

An enormous shagbark hickory guarded the number six green on the Blackhawk Country Club Golf Course in Madison until the late 1990s, when it was removed because it had died. The location is still marked by a handsome brass marker set into a granite boulder taken from the terminal moraine at Cross Plains. Credit is given to Walter E. Scott, who discovered and recorded the tree's size. Several Indian effigy and burial mounds are also marked on this historic site bordering Lake Mendota in the village of Shorewood Hills, which has been recognized for its importance as a place frequented by Indians in the early days.

A prehistoric Indian group called the Effigy Mound Builders constructed the mounds in the period from 500 to 1500 A.D. Wisconsin once had a treasury of these mounds, but the early settlers, not being very interested in the odd ridges or knolls, destroyed most of them by plowing or excavation. Increase Lapham, scientist and archeologist, was the first to identify the mounds and recognize their importance. He reported on them in an article titled, "The Antiquities of Wisconsin," published in Smithsonian Contributions to Knowledge in 1855. The *Wisconsin Archeologist* for 1910 also discussed the mounds.

Because of these well-preserved Indian relics, the Blackhawk Country Club was named to the National Register of Historic Places and was declared a National Historic Site by the National Park Service in 1979.

For many years the shagbark hickory stood in distinguished company. Its large, twin-forked trunk, 11 feet 5 inches in circumference at breast height, made it a majestic young companion to its ancient neighbors: the flying goose, the three bears, the panther, and the hawk. Today, a bike path passes through where the Blackhawk hickory once stood.

Blackhawk Hickory

B-Wolfgang Hoffmann (1979)

Horicon Marsh Trail Marker Tree

Edgar G. Mueller (1971)

Trail Marker Trees

It's hard to imagine Wisconsin, a tourist state, without its ubiquitous highway signs of every shape, size, and description. But even early residents of the region needed help to get where they were going.

Trees marked trails and pointed out river fords, burial grounds and other places of interest. There is debate today among archeologists as to whether Indians or early settlers actually manipulated trees to cause them to grow into unusual shapes pointing toward landmarks. But even without scientific evidence of accuracy, fascinating legends have developed around these unusually shaped trees.

An article in the *Wisconsin Archeologist* in 1965 by Phil Sander, a naturalist and conservationist, suggested that most permanent marker trees were made from hardwoods such as oak, hickory, maple and elm, because of those trees' flexibility and strength. According to Sander, the tip of a sapling might be buried in the ground or tied to a stake by twisted vines or strips of hide. As the tree continued to grow, the tip would die off and the trunk would assume a bent position. Usually a new upright trunk formed from the old tip. As the tree matured it took shape as a "living road sign that could be easily recognized and would last the life of the tree."

A distinctive tree was found at Twin Lakes in Kenosha County near the top of a high ridge overlooking the northeast bay of the lower lake. Oriented south, it pointed toward the land bridge separating the two lakes and toward the site of an old village and cemetery. Sander discovered another unique tree in Adams County in 1977. Located in Springville Township, the tree was a few miles east of the Wisconsin River bridge on the south side of Highway 82 in a picnic area. It pointed west.

At Green Lake, on the Victor Lawson estate, trees indicated an old trail passing through those grounds.

Two fine examples of trees that may have marked trails were found in the Horicon Marsh area. One, at the west end of Horicon, was a large oak with a massive branch pointing in one direction. Just west of Highway 28 on a farm was a large bur oak, possibly

Twin Lakes Trail Marker Tree discovered by Phil Sander (1965)

upwards of 200 years old, with interesting branches twisting in several directions.

Southeast of Brodhead, near reconstructed Highway 81 alongside Skinner Road, stood a white oak that dated back to pre-settlement days. Its odd bent shape suggested that the tip of the tree had been tied at ground level and a new terminal shoot had grown up at right angles to it. According to some stories, this tree helped Indian runners mark the trail from the Great Lakes to the Mississippi River, a trail that is now followed by cars on Highway 81 between Beloit and Monroe. Myron Olson, who grew up near Skinner Road, recalls playing on the tree as a youngster. His grandfather operated a blacksmith shop nearby. Another tree thought to be a marker tree, at Bluff View Park, came down some years ago when a new road went through.

An oak in the yard of the Walter Atwood home in Indian Hills, Madison, was also thought to be a trail marker tree. Now just a short distance from the golf course at Blackhawk Country Club, this tree may have marked the trail Black Hawk and his followers used to flee towards the Wisconsin River in 1832.

Mercer's Addition Trail Marker Tree Wisconsin Historical Society, WHi(X3)38163

A possible trail marker tree once stood in what was then called Mercer's Addition in Madison, near West High School. A hickory, it marked the crossing of two trails, and its branches, rather than its trunk, appeared twisted to point in four directions.

Another vanished tree is a beech that once stood near the corner of Wells and Thirteenth Streets in Milwaukee. Into the trunk was cut an Indian figure with a bow in one hand and an arrow in the other. The arrow pointed to the Menominee River.

Sources: Mrs. Walter Atwood, Madison
Dorothy Gessert, Brodhead
Phil Sander, Kenosha

Chief Tokaunee Oak B-Wolfgang Hoffmann (1982)

The Chief Tokaunee Oak

Dane County

A bur oak growing along Highway 19 near Token Creek in Dane County has a special distinction. A family loved it enough to buy it. Originally Dr. and Mrs. John Atlee's property was adjacent to but did not include the tree. The family bought the land it grew on because of the old oak's beauty and climbability, and because they wished to preserve it from encroachment. They knew it had watched the comings and goings of both Native and white Americans for the past 200 years.

According to the book *Dane County Place-Names* by Frederic G. Cassidy, Token Creek and the settlement by it were probably named for Tokaunee, a mixed-blood chief of the Winnebago tribe, which frequented the area during the time settlers were beginning to arrive in the territory. The village of Token Creek, though never platted, was one of the first settlements in Dane County and gained considerable importance during its early years.

Sources: Dr. and Mrs. John Atlee III, Token Creek

Chapter VI

Surveyor's Trees

Whitewater Territorial Oak Rolland P. Schlieve (1981)

The Territorial Oak

Whitewater

A huge old bur oak tree, from which early surveyors platted the village of Whitewater, has been adopted as the symbol of that city. It is pictured on the official city stationery and also appears in the famous Whitewater mural located at Whitewater and Center Streets. On February 17, 1981, the Whitewater City Council nominated the Territorial Oak for consideration as a historic tree.

The handsome old tree, which stands on the corner of Franklin and Main Streets, graces a home that once belonged to Rosetha Trippe, one of Whitewater's earliest settlers. Rosetha came from New York state with her husband, Dr. James Trippe, in 1836. Dr. Trippe proceeded to build the first grist mill on Whitewater Creek. As the only one in the vicinity, the mill was patronized by farmers from a large area, contributing substantially to the prosperity of the fledgling city.

The Trippe's first home, in which Rosetha entertained all visitors to the area, was a log cabin near the mill site. It was not until much later, after Dr. Trippe died, that Rosetha moved into a house built on the corner lot shaded by the Territorial Oak.

Sources: Mr. and Mrs. Alfred Kolmos, Whitewater
Quinn C. Smet, Whitewater

The Cram/Houghton Blaze Tree

Vilas County

The first white people to set foot in Wisconsin weren't all explorers, fur traders, soldiers or lumberjacks, although it is often these types about whom tales are told. A great many were surveyors who had the responsibility of marking the boundary lines for the entire territory, a necessary prerequisite for statehood.

Thomas Jefferson Cram was such a surveyor, and he came to Wisconsin Territory in 1840 on a mission to check an earlier surveyor's work: the northeastern boundary, as described in the enabling act for Michigan. According to Cram's reconnaissance report to the War Department, the previous survey line was inadequate. In March 1841, Congress ordered a second survey and Cram came again to Wisconsin to carry it out, bringing with him Douglass Houghton, later to become the well known engineer for whom the city of Houghton, Michigan, was named.

They had an incredible task: surveying a boundary over 100 miles long. Both men endured great hardships during the work. Several years later a souvenir of their efforts was discovered. On a large pine tree growing on the shore of Trout Lake in Vilas County, on a blaze two feet long and one foot wide, had been cut the words: "XIX T. J. Cram D. Houghton Aug. 11, 1841."

Unfortunately, lumbermen did not spare the tree, but the slab with the inscription was saved and presented to the Wisconsin Historical Society by John B. Mann of Minocqua.

Cram/Houghton Blaze Tree in the Wisconsin Historical Museum

B-Wolfgang Hoffmann (1982)

Witness Trees

When surveyors came to Wisconsin in the early 1830s and '40s to survey the territory prior to statehood, they often sought out trees as markers or "bearers." In the south, bur oaks, already growing on the prairie because they were resistant to fire, were most often used. Some of these trees were large landmarks even then. At other times, the surveyors bent young trees into abnormal shapes to mark the property lines. There are undoubtedly many of these trees still standing, but we know about only a few of them. For example, Dunn, St. Croix, Jackson, Jefferson, Waushara and Dane Counties have all reported finding identifiable witness trees.

Resurveys of property that necessitated finding the old bearing trees have resulted in the entering of some updated descriptions into the official records. From Oakland Township in Jefferson County are two such references:

> *Found Original Bearing Tree, measured distance & angle to the corner dug down 1' and found the Stone ... Tree age by boring is over 250 years as of date below (1975) and in good shape....*
>
> *A very large Bur Oak fits original Bearing description, but was unable to see any markings on the tree, Bearing would be to the Southeast of the concrete marker.*

A rare bearing tree with a sideways "A" carved into a blaze on the trunk stood on the property of an old-time logger, J. M. Rustad. Its location marked the survey line between Albany and New Hope Townships in Portage County.

In the southwest corner of Section 30 in the Town of Richford in Waushara County, a black oak served as a witness tree. While resurveying around 1980, the county surveyor found the measurements precise to the old tree, which had served as the section corner point since about 1850.

Sources: Mrs. Hugh F. Gwin, Hudson
Ray H. Hansen, Coloma
Lawrence Jones, Black River Falls
Hannah Swart, Fort Atkinson

Hanerville Oak

Norris Linderud (1955)

The Hanerville Oak

A large handsome bur oak tree, called the Hanerville Oak, used to loom like an impregnable traffic cop in the very center of an intersection, halfway between Stoughton and Edgerton in the Town of Dunkirk in Dane County. It stood across from the old stone Hanerville School, now closed, and the children who went there accepted the tree matter-of-factly, not realizing the effort it took to prevent the tree's destruction in 1934 when County Highway N was being rebuilt. The tree grew in the center of the proposed road. A man named Tharon Miller led the fight to save the oak because it had been a witness tree, a section marker used by surveyors in 1833. After all, the oak had seen the early settlers pass beneath it on their way from Janesville to Madison, and it had no doubt noted the arrival of Luke Stoughton as he neared the site of the town he would found.

The Edgerton Chapter of the Daughters of the American Revolution agreed with Miller. Their sentiment prevailed, and the DAR marked the tree with a bronze tablet that called attention to its historic importance.

Now, unfortunately, the tree is no longer standing. While this means that motorists unfamiliar with the area will not need to worry about bumping into a tree growing in the middle of an intersection, it also means that they will never be made aware of an interesting bit of history along the route.

Source: Norris Linderud, Hanerville

Chapter VII

Ancient, Huge and Unusual Trees

Washington Island Pine

December 1955 marked the end for one of the largest white pines in Wisconsin, a 450-year-old giant that stood on a knoll on the east side of Washington Island. Approximately 150 feet high and about 15 feet in circumference, it must have been a seedling when Columbus discovered America.

The pine was a youngster when the Indians still roamed in peace and plenty under its branches. It witnessed the coming of the white settlers: the Icelandic Colony, the Scandinavians, later the Irish and Scots. Most of its fellow pines were felled for lumber, but, because it was in an inaccessible spot, it remained, coming to stand virtually alone.

Measured and treasured, the Big White Pine was for years a prime attraction for visitors to Washington Island. This was possible because the owner of the land on which it stood, Arnold Wickham, appreciated its age and nobility. Though some advised him to cut it down for the thousands of board feet of lumber it contained, he refused. Instead he cleared a road to the monarch so that others could be inspired by it. Because of occasional souvenir seekers who took cuttings from the tree, he put up a fence.

After it was fatally damaged by a lightning strike, professional woodsmen came to cut down the Big White Pine in what was one of the largest logging operations on Washington Island in 65 years. Wallace Jensen, chief lumberjack, directing his crew of handpicked men, used special equipment to fell it. The logs were moved out of the woods on a specially constructed skid pulled by a cat tractor. Taken to the sawmill, the lordly monarch became merely lumber. Mr. Jensen remarked that it was a thrill watching those wide pine boards come off the saw carriage.

Source: Conan Bryant Eaton, Washington Island

The Phelps Oak

Dodge County

On Wednesday, September 11, 1964, high winds buffeted Dodge County. Among the casualties was one of the largest bur oaks in the state, a mammoth growing on the farm of H. M. Phelps on State Highway 33, about three miles west of Fox Lake.

At one time Walter Scott, record tree recorder, reported it as the largest bur oak in the state, measuring 14 feet 1 inch in circumference, with a height of 88 feet and a crown spread of 108 feet. Due to Scott's efforts, however, people became more aware of the record-size trees around them, and the Phelps oak eventually lost its top ranking.

Still, the Phelps oak enjoyed a deserved fame, making the cover of the *Wisconsin Conservation Bulletin* in October 1952 and serving as a bench mark for its species. It was hollow when it blew down, prompting a dramatic photo in the Beaver Dam *Daily Citizen*. When such a monarch crashes to earth, it does not go unnoticed.

Phelps Oak — *Wisconsin Agriculture and Farmer* (c. 1952)

Pine Street Oak

B-Wolfgang Hoffmann (1982)

The Pine Street Oak

Madison

One of the oldest trees in Madison is a white oak that dominates its corner lot at 602 Pine Street on Madison's south side.

The pleasant old neighborhood, beautified by stately oaks and elms and featuring streets named Spruce and Cedar as well as Pine, was first entered at the land office in 1836. It was platted August 8, 1890. Called Wingra Village, because Wingra Creek forms one boundary, it used to be considered a suburban location. The first inhabitants represented a diverse variety of ethnic groups, and many well-known Madison names were once listed in its directory. They include Henry Turville and Frank Custer, Madison's historian extraordinaire who is honored by his own tree, the Custer Ash, planted in James Madison Park by the Daughters of the American Revolution (DAR) on Arbor Day, 1979. The certificate cites him for "making Madison history come alive."

In the late 1970s, R. Bruce Allison was called to stop a neighbor from felling a nearby tree that would have crashed directly across several limbs of the venerable Pine Street Oak.

At that time the house on the huge white oak's lot had been occupied by the same family for over 50 years. They had a love and concern for the tree's well-being as if it were a neighbor and friend. The family's generosity of spirit was well expressed when they tied a long yellow ribbon around the oak's trunk in celebration of the return of the U.S. hostages from Iran in January 1981.

Sources: Frank Custer
Mrs. Duane K. Smith

Maple Bluff White Ash B-Wolfgang Hoffmann (1979)

The Champion White Ash

Maple Bluff

A fierce windstorm in 1981 felled one of the most magnificent trees in the Maple Bluff section of Madison. A majestic white ash, it held the state record as the largest of that species so far recorded in Wisconsin. Its trunk measured 13 feet in circumference; it was 79 feet high and had a spread of 80 feet.

The Duane Bredesons, who lived in the house, had always worried that the giant, if hit by wind or lightning, might split their home in two. But the ash did no damage as it died. Inside the house it is much lighter—and in the summer, hotter—now that the tree is gone.

Before the European explorers came, the ash towered above the Indians who roamed the lake shores. Many of its fellow trees were sugar maples, which later gave their name to the promontory first known as Sugar Grove. On an old surveyor's map, used from 1831 to 1837, a field note reads: "Here is located a large grove of sugar trees of about 200 acres."

Maple Bluff is still noteworthy for its beautiful old trees.

Source: Mrs. D. H. Bredeson, Maple Bluff

The Montello Cottonwood

From a standpoint of size alone, one of the most famous and magnificent trees in Wisconsin is the Montello or Troost cottonwood on Highway 23 at Park Street in Montello, Marquette County. A perfect, symmetrical specimen, it is one of the largest in the state.

Although cottonwoods are a fast-growing tree, this giant is probably 200 to 300 years old, which means that it was standing when the French voyageurs passed through and gave the name "Mont L'Eau" to the area. It was certainly large when the first settlers came in 1844. It grows on property owned by Mrs. Helen Troost, just a block from the downtown section.

To those who live or work near the tree, it is a thing of majestic beauty through the seasons. One of its staunchest fans was Robert L. Wright, who edited the *Marquette County Tribune* for 30 years. The tree was visible from his office window, and he grew to love it. In his own words: "If some day a summer storm brings it to earth, we who have lived with it will be poorer. For 30 years, I labored in my newspaper shop directly across the street from the Troost cottonwood and never a day went by that I did not look and admire. The tree was—and is—my friend."

Sources: J. A. Kronschnabel, Montello
Robert Wright, Montello

Montello Cottonwood

Mary Frances Schjonberg (1982)

Columbus Cottonwood

B-Wolfgang Hoffmann (1979)

The Columbus Cottonwood

Dodge County

An Eastern poplar (cottonwood) reigned over the John Crombie farm near Columbus in Dodge County as the tree with the largest trunk circumference in Wisconsin. Its trunk, which divided part way up, measured 26 feet 2 inches in circumference; it was 99 feet tall with a crown spread of 101 feet in 1980 when it was measured for inclusion in the first edition of *Wisconsin's Champion Trees*. In total points combining circumference, height and crown spread, however, it fell short of the Montello cottonwood, which, though smaller in girth, was taller and had a bigger crown spread. Remeasured for the 2005 edition of *Wisconsin's Champion Trees,* the Columbus cottonwood circumference has grown to 30 feet.

Although the Crombie family had lived on the farm for two generations, John Crombie says he didn't remember his father, who bought the farm in the 1930s, ever mentioning the tree. It stood about 50 or 60 rods from the house and was visible from the road, but somehow it didn't look as large as it was. It was only when Crombie read in the Beaver Dam *Daily Citizen* about another tree's claim to the top rank that he went out and measured his own, sending the figures to the tree record compilers at the Wisconsin Department of Natural Resources.

For more than a century, this top contender remained modestly in the background, content with doing the things trees do best. With the publication of *Wisconsin's Champion Trees* in 1980, however, it gained a deserved fame as a cover tree and poster.

Source: John Crombie, Columbus

Two-Leader Pine

Eau Claire County

Many Wisconsinites like to reminisce about the days when the northern section of the state was virgin pine. The logging boom of the nineteenth century left very few of those original trees. But there was one relic that escaped the woodsman's ax in Eau Claire County. It grew about two miles south of the city of Eau Claire, and foresters believe the tree was spared because of the difficulty of getting it out of its inaccessible site.

In 1950 the tree was almost 14 feet in circumference. The odd thing about the tree was its twin tops. Apparently the leader had been broken off some time in the past and a lateral branch had sent up a new terminal shoot. A strange-looking tree indeed. The owner at the time, appreciating its novelty, agreed to preserve it. It is unknown whether this tree still stands.

Two-Leader Pine Jim Fisher

The Stockbridge Cottonwoods

Calumet County

We have all experienced the feeling, when we revisit the scenes of our childhood, that things are a lot smaller than we remembered them. Yet a man who lived all his life with two giant cottonwood trees in his front yard said the trees seemed just as large to him at age 67 as they did when he was a boy. Proof, perhaps, that living trees are one of the few things that can keep up with a person's changing perceptions of size.

The two trees stood a couple of miles south of Stockbridge and a half-mile west of Highway 55 in Calumet County. The one in front of a small cemetery, shading the gravestones almost entirely from the morning sun, was the second largest cottonwood in the state at one time. Its companion tree, almost as large, stood in a farmyard a short distance away. Both trees were just about half a mile from the eastern shore of Lake Winnebago. Perhaps they were planted by the Stockbridge and Brothertun Indians, remnants of New England tribes that lived in New York State before moving to Wisconsin about 1834.

The Black Tupelo Tree

Kenosha County

A very rare black tupelo tree inspired a bit of detective work on the part of two naturalists who finally tracked it down. Early botanists had recorded two such trees in 1909 in Somers Woodlot, located in Somers Township, Kenosha County. Apparently this small wooded area contained a community of about eight species of hardwoods. Yet it was not until May 1965, 56 years later, that the naturalists rediscovered a single tupelo tree (sometimes called sour gum, black gum or pepperidge).

In the 1909 reference work, the author of "Flora of Racine and Kenosha Counties" reported the tupelo trees growing "within a stone's throw of the P.O. at Berryville, Kenosha County ... the only trees of their kind known in the state." Botanists were interested because proof of this rare tree's existence would extend considerably its known northern range.

It took several tries for Phil Sander and Henry Otterson, both of Kenosha, to find the site of the old woodlot, but the two large black tupelo trees were gone. They believed the single young tree to be an offshoot. It was growing on the north side of Sixth Place and Sheridan Road in Berryville. In the late 1970s, botanists discovered several younger tupelo trees in the same woodlot. An interesting characteristic of these trees is the oval, berry-like drupes, about a third of an inch long and blue-black when ripe.

Adding to the mystery of the tupelos is the speculation about how they came to be growing in Wisconsin. One theory holds that perhaps the seeds of the trees were transported in the crops of the once-numerous passenger pigeons as they migrated from southern climes through Wisconsin, using the nearby Pike River Valley flyway and feeding grounds featuring their favorite food, acorns.

Source: Phil Sander, Kenosha

Kettle Moraine Scary Oak Elizabeth Deakman

The Scary Oak

Kettle Moraine State Park

There's a downright spooky tree growing along the nature trail in the Kettle Moraine State Park in Walworth County. It is an old bur oak, monstrous in size and appearance. Typically, open-grown bur oaks develop short, heavy trunks to support wide-spreading crowns composed of massive lateral branches. These lateral branches and twigs grow crooked, giving the tree a corkscrew appearance.

The sign in front of the "big, bumpy, burly, bur oak" explains that burls—a kind of tumor-like, uncontrolled growth of wood cells—are often started by insect damage, fungus disease, or physical abrasions. Burls tend to make a tree useless as timber, and they certainly make it impossible to split up into firewood. But the tough and wavy-grained burls are sought after by wood artists, for they can be made into magnificent bowls and other art or household objects.

Source: Elizabeth Deakman, Madison

The General MacArthur White Pine

Forest County

Although General Douglas MacArthur never saw this tree, at one time it was probably one of the largest known white pines in the United States or the world. It was certainly a magnificent relic of the past and deserved to be one of the attractions that lured thousands of tourists annually to the Nicolet National Forest, near Newald in Forest County.

Estimated at over 400 years of age, and given the name of one of Wisconsin's best-known citizens in honor of both of them, the MacArthur Pine, with a circumference of 17 feet 6 inches, height of 140 feet, and crown spread of 48 feet, ranked as the second largest white pine in Wisconsin in the 1980s.

The towering sentinel gave us an indication of what the northern portion of the state must have looked like before the lumberjacks got to work with their axes to make Wisconsin, by 1907, pre-eminent in lumbering among all the states. It is figured that the tree contained more than 8,000 board feet of lumber, enough to build a house.

General Douglas MacArthur White Pine DNR (c. 1946)

The tree that Paul Bunyan must have spared was discovered by a U.S. Forest Ranger in 1945. It could be seen about four miles from Newald beside Highway 139.

Erosion over the centuries uncovered the pine's huge, gnarled roots and left them hollowed and open to small burrowing animals who shared the tamarack swamp with the giant. Unfortunately, root disease and decay and fires from lightning strikes took their toll. The MacArthur pine burned to the ground in 2003, possibly due to lightning.

Standing at the foot of the tree and looking straight up for 80 feet to the first branches, one could see why shipbuilders prized *Pinus strobus* for the masts of sailing vessels. The MacArthur Pine was indeed a noble specimen, better known and more photographed than the state's largest pine (larger only by inches) in the Brule River State Forest in Douglas County.

Source: Bob Ellingson, Lakewood

Brule River White Pine

Douglas County

For many years, the state record white pine, whose companions were endangered eagles and ospreys, had a protected habitat in the Brule River State Forest. In the early 1980s, The Nature Conservancy had acquired by donation easements that effectively shielded 4,800 acres of privately owned land along nine miles of the Brule River from further encroachment. In addition to the former record holding white pine, the easement area contains one of Wisconsin's largest and highest quality virgin pine forests, with many individual trees approaching 300 years in age. Also protected is one of the state's largest cedar swamps, containing an endangered orchid and several unusually rich and diverse forest habitats, supporting more than 90 bird species, many of them quite rare in Wisconsin.

In 1962 the Brule white pine had a circumference of 17 feet 11 inches, only seven inches short of the national record white pine in Michigan.

As have several presidents, Calvin Coolidge spent some time along the Brule River in 1926, fishing and establishing a summer White House at a well-known resort called Cedar Island Estate. (During his stay in Wisconsin, he located the graves of his grandparents,

Brule River White Pine DNR (c. 1962)

Israel and Sally Brewer, in Columbia County.) One can imagine the 30th president of the United States, tiny and insignificant in size, standing beneath a record pine and gazing up that 151 feet of trunk, no doubt feeling a kinship with the old monarch in the practice of silence.

In the early 1980s one of the three large limbs of the pine fell, narrowly missing employees of the fish hatchery below. Unfortunately, even after pruning and repair by a volunteer crew of Allison Tree Care, Inc., it succumbed to old age and died.

Source: Russell Van Herik, Madison

The Lunde Chestnut Trees

Trempealeau County

The American chestnut, undoubtedly one of the nation's most loved species of tree, enhanced by folklore, nostalgia and literary allusion, is commonly thought to be as extinct as the dodo because of the chestnut blight introduced from Asia in the very early years of the twentieth century.

It is true that most of the country's chestnut stands, certainly those in the eastern part of the country, were completely wiped out before 1940. This total destruction of a native tree species was something never witnessed before that time. But contrary to what most folks believe, chestnuts are alive and living in Wisconsin. They make up a significant portion of the genetically pure American chestnuts left in the world.

A large grove of them, many nearing the hundred-year-old mark, grows in western Wisconsin, on the farm of Einar Lunde, three and a half miles east of Trempealeau in Trempealeau County. There are also small groves of fruiting chestnut trees in Green Lake and Sauk Counties, and single chestnut trees are scattered from Kenosha to Bayfield.

These Wisconsin trees are outside the chestnut's native zone of 20 eastern states. During the years of the blight's scourge, they were protected by the effective barrier of 500 miles of open country. Prevailing westerly winds also helped keep the fast-acting and fatal fungus, for which no cure is presently known, out of Wisconsin until recently.

The trees on Einar Lunde's farm provide more than just esthetic enjoyment, though the sight of them covered with snowy white catkins about the first of July is a thrilling one. Their copious harvest of nuts is also gathered, both for the delicious eating and for their use by forest researchers in efforts to breed a blight-resistant chestnut.

Chestnut trees have received frequent mention in the annals of Wisconsin history. Increase Lapham, the state's first scientist, recommended over a hundred years ago that chestnuts be planted because of the value of their nuts, favored by humans and beast alike. There is evidence that early settlers brought the chestnut to Wisconsin because they cherished not only its fruit and the beauty of its flowers, but also its all-round usefulness for fence posts, shingles, poles and ties, charcoal, etc.

In 1859, it is recorded, a farmer named Jacob Lowe, from Columbia County, won a prize from the State Agricultural Society for the first chestnuts produced by his young trees.

Wilmot Camperdown Elms Phil Sander (1963)

The Upside-Down Trees

Wilmot

People used to believe that the two weeping elms in a front yard in Wilmot in Kenosha County, and others like them, had been planted upside down. These camperdown elms *(Ulmus glabra camperdownii)* have peculiarly drooping, vine-like branches and flattish tops. They do indeed look as if the roots, sticking up in the air, had put out leaves. This popular misconception was widespread. As is true of many folk beliefs, the myth is probably more interesting than the reality.

The camperdown elm, a mutant, sometimes called the European mountain elm, was first developed from an original plant which grew at Camperdown House near Dundee, Scotland. A vine-like elm was grafted on top of an upright elm in order to produce the tree. In the Wilmot elms, as well as others of the type, the graft is still visible as a swollen ring about halfway up the trunk of the tree.

The "upside-down trees" were popular ornamentals around the turn of the century and still grace the lawns of old mansions and public buildings. Actually, of course it is botanically impossible for roots to bear leaves.

Sources: Pamela Fullerton, Madison
Phil Sander, Kenosha
Gen Turk, Kenosha

The Buried Forest

Kenosha

There's a buried forest just south of the city of Kenosha. Along the shore of Lake Michigan, in the area known as the Southport Dunes, shifting waves eroded the sand to reveal evidence of a deciduous forest that thrived there over 6,000 years ago. Roots and branches of trees, uncovered by the violent wave action, have been carbon-dated to be approximately 6,340 years old.

Four stumps embedded in a 12-inch layer of black humus overlying a stratum of clay were exposed in 1961. The tree remnants, according to experts from the University of Wisconsin, the Forest Products Laboratory, the Kenosha and Milwaukee Museums and the U.S. Geological Survey, are identifiable as three red oaks and one elm. The diameters of the ancient trunks ranged from 12 to 15 inches. The wood is black and spongy. Phil Sander, a Kenosha conservationist and naturalist who has made several discoveries along the beach, says the trees probably grew during the last interglacial stage and were reburied during a temporary re-advance of the last ice sheet. The shifting sands are constantly re-covering the ancient stumps as they reveal other roots and branches, some identified as ash and oak.

The study of the buried trees is important for several reasons. The types of pollen still in the logs can tell us about the kind of vegetation abundant during the Pleistocene geological epoch, as well as the date deciduous trees began to establish themselves in Wisconsin.

Source: Phil Sander, Kenosha

Chapter VIII

Arboretums, Campus Trees and Forests

Muir Locust

University of Wisconsin–Extension (1953)

The Muir Locust

Madison

It was beneath a black locust tree, which stood along Observatory Drive near North Hall (the first building) on the University of Wisconsin–Madison campus, that John Muir, famous naturalist and father of the national park system, had his first botany lesson from a fellow student.

Muir, a Scottish lad from a farm near Portage, had come at age 19 to the new University seeking a college education, though his previous schooling had been slight. A fellow student took Muir to observe the blooms on the locust tree, explaining to the fascinated farm boy some of the rudiments of botany. It was June 1863. Here is Muir's own account of the incident from his autobiography, *The Story of My Boyhood and Youth:*

> *I received my first lesson in botany from a student by the name of Griswold, who is now County Judge of the County of Waukesha, Wisconsin. In the University he was often laughed at on account of his anxiety to instruct others, and his frequently saying with a fine emphasis, "Imparting instruction is my greatest enjoyment." One memorable day in June, when I was standing on the stone steps of the north dormitory, Mr. Griswold joined me and at once began to teach. He reached up, plucked a flower from an overspreading branch of a locust tree, and, handing it to me, said, "Muir, do you know what family this tree belongs to?"*
>
> *"No," I said, "I don't know anything about botany."*
>
> *"Well, no matter," said he, "what is it like?"*
>
> *"It's like a pea flower," I replied.*
>
> *"That's right. You're right," he said, "it belongs to the Pea Family."*
>
> *"But how can that be," I objected, "when the pea is a weak, clinging, straggling herb, and the locust is a big, thorny hardwood tree?"*

Griswold then introduced to the receptive student the principals of taxonomy, the classification of all plants by shared characteristics. Muir's eyes were forever opened to the remarkable orderly world of botany. The big, thorny hardwood tree, relative to the weak clinging pea, in tribute to Muir's stature, as a protector of the natural world was later named the Muir Locust.

Muir Locust University of Wisconsin–Extension (1955)

Muir died in 1914. In 1918 the Board of Regents dedicated Muir Knoll, across Observatory Drive from the famed tree. A red granite boulder marks the site. Charles H. Vilas delivered the dedication address, and Muir's "teacher," Judge Griswold, and roommate Charles F. Vroman spoke of their recollections of the great naturalist.

Unfortunately, in 1953, amid considerable controversy, the University cut down the tree, by then 12 feet 1 inch in girth. Many, including Walter Scott and Wakeland McNeel, argued forcefully, yet futilely, to save it. The justification provided for its removal by G. William Longenecker, professor of horticulture and landscape architect, was that the tree's heart was dead and its trunk filled with pitch, and that further efforts to save the tree would be useless as it had already lived many years past its prime. Professor Walter Rowlands made the wood into gavels, letter opener knives and other similar objects. Then President E. B. Fred passed these souvenirs on to friends of the University who had some contact with the Muir legacy.

In researching this story R. Bruce Allison recovered long forgotten, damaged photographs of the tree and its removal stored as negatives at the UW photo laboratory. On a hunch, he delved into the dust-covered files of the late E. B. Fred stored in the lower basement archives at Memorial Library. Looking for thank you note responses to the souvenir gifts, he came across one particularly poignant letter that told the story of the conflict of the tree's removal.

John Muir
Wisconsin Historical Society, WHi(X3)5766

In response to receiving one of the gavels, Wakeland McNeel, then supervisor of the Upham Woods Project for the University and who was also known as Ranger Mac on public radio, wrote:

> *I do not know what we are going to do now that the tree is gone. It stood there as a tangible, effective, though silent spokesman of a turning point in a man's life; of a time of decision that made the outdoors his University and gave the world John Muir.*
>
> *A gavel shouts for order. This tree, in the grand, silent manner of its waning strength, secured order like prayer. To stand beside the distorted tree body, to pat its wrinkled skin, to throw your arms about its scarred body, and then to tell the story, was a memorable way to impress young folks gathered about that there lived a man like John Muir, and that trees can do divine things.*
>
> *I do not know now what we are going to do to accomplish these results. But I feel highly pleased that you remembered me with a portion of the flesh of this tree, shaped in the beautiful form of the gavel. I have placed the gavel on the shelf along with his books with the silent prayer that the memory of the accomplishments of this great man, and the accomplishments themselves, will never go the way this tree had to go.*
>
> *It was most gracious of you to remember how deeply I hold in reverence this foster son of our state.*

Muir Park, including the knoll and about seven acres formerly known as Bascom Woods, was dedicated on February 8, 1964, as a laboratory for the study of flora and fauna of the region. The speaker, Professor Grant Cottam of the botany department, described the spot as "One of Nature's history books ... representing a thousand years of vegetational history."

Muir himself left the University after four years without actually receiving a degree. Later he wrote in his autobiography:

> *From the top of a hill on the north side of Lake Mendota I gained a last wistful, lingering view of the beautiful University grounds and buildings where I had spent so many hungry and happy and hopeful days. There with streaming eyes I bade my blessed alma mater farewell. But I was only leaving one University for another—the University of the Wilderness.*

Source: Frank Cook, Madison

The Harvey Oak

Madison

Until the 1990s, when it was taken down because of defects, a black oak tree stood at the southwest corner of Camp Randall in Madison with a plaque inscribed: "Dedicated to the memory of Governor Harvey."

Louis Powell Harvey became governor of the state in 1862. He was the second Civil War governor (Randall was the first). Harvey had a deep concern for Wisconsin soldiers wounded at the Battle of Shiloh. His compassion led him to organize a journey to the military camps on the southern rivers where the Wisconsin men were being cared for in makeshift hospitals.

Harvey Oak Mary Frances Schjonberg (1976)

Crossing from one steamboat to another in the night, he lost his footing and was drowned in the Tennessee River. His body was finally discovered more than 60 miles downstream.

His wife, Cordelia, hoping to carry on her husband's work, was appointed sanitary agent to the Union Army. She visited all the hospitals along the Mississippi River. The shocking conditions led her to collect signatures on a petition that she carried to President Lincoln in Washington. Her hope was to have hospitals established in the healthier climate of the North. After long debate with Lincoln, she succeeded in getting her wish. The first such hospital was established in Madison, in an octagonal building on Lake Monona.

Perhaps the black oak should have had a dedication to Cordelia Harvey, too.

The President's Tree

Madison

A large bur oak near the old astronomer's residence on Observatory Hill, on the Madison Campus of the University of Wisconsin, is known as "The President's Tree." It has been there longer than the University itself. The house formerly served as home to a succession of University presidents.

The tree is massive and is said by botanists to be over 250 years old.

Legend has it that during the Civil War, soldiers stationed at Camp Randall used the tree for gunnery practice. A gaping hole in the trunk was offered as evidence. As protection for the beloved tree, President E. B. Fred had it bolted together with long steel rods. It appears to be in perfect health, possibly good for another 250 years.

President's Tree

B-Wolfgang Hoffmann (1982)

The Bascom Hill Elms

Madison

Think of Bascom Hill and you remember the great elms. The pinnacle of the University of Wisconsin–Madison campus was famous for its parallel, tree-lined walks. Some of these majestic trees, going back to the earliest days of the University—the 1850s—have died from Dutch elm disease. Many of those treated with fungicide for this disease have survived.

As of 2004, 34 mature elms remain on campus, with 16 of them on Bascom Hill. In spite of the fact that they are receiving the best of care, chances are they will all eventually disappear. Meanwhile, primarily red oaks are being planted there as replacements. They will, some day, become the dominant trees on Bascom Hill. But will they be able to replace the memory of the elms?

Bascom Hill Elms

B-Wolfgang Hoffmann (1980)

Willow Drive Willows (above and below)

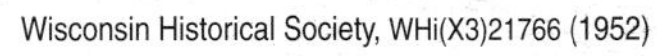

Wisconsin Historical Society, WHi(X3)21766 (1952)

B-Wolfgang Hoffmann (1982)

The Willow Drive Willows

Madison

At the end of North Livingston Street on Lake Mendota were willows said to have come from the grave of Napoleon on St. Helena. Other willows made a tunneled roadway of Willow Drive along the lake on the UW–Madison campus. These trees were planted under the direction of President Thomas Chamberlain with the assistance of Professor John Olin in the 1880s.

Road expansion in 2004 required removing most of the willows growing between Picnic Point and the Boat Launch. Thirty-eight willows along with river birch and swamp white oaks, for diversity, have been planted so that the famed Willow Drive will continue to be one of the most beautiful drives in Madison.

Source: Daniel Einstein, Madison

Geotropic Goff Larch B-Wolfgang Hoffmann (1982)

The Geotropic Goff Larch

Madison

While he was president of the University of Wisconsin–Madison, and until he died in 1981, E. B. Fred was fond of a large old larch tree that, strangely, bent its branches down instead of up. The European conifer is at 10 Babcock Drive, traditional home of deans of the College of Agriculture.

The larch was brought from Door County when it was only two feet tall, in 1899, and was dedicated to Emmett S. Goff, UW professor of horticulture. Landscape architects have tried many times to get cuttings from the larch to grow in the same way as the parent tree, with branches pointing down. The cuttings refuse, preferring, apparently, to grow normally. No one knows why.

Autumn Purple White Ash B-Wolfgang Hoffmann (1980)

The Autumn Purple White Ash

Madison

The very first known specimen of the Autumn Purple White Ash, now a registered trademark for a nationally distributed tree, stood to the southwest of the Home Economics Building on the University of Wisconsin–Madison campus until it was cut down in the 1990s because of structural problems.

G. William Longenecker, a well-known professor of horticulture at the University, recognized this staminate tree for its good form and spectacular autumn foliage. Longenecker worked with McKay Nursery of Wisconsin and Cole Nursery of Ohio to trademark and make available this official cultivar. It is widely appreciated all over the United States.

The first sapling propagated from the campus tree was planted in the Longenecker Gardens at the UW Arboretum, and the second at the Boerner Botanical Gardens in Hales Corners. Unfortunately, there is no record of who originally planted the tree on the campus or where it came from.

The Euthenics Oak

Madison

Below the UW's Human Ecology Building west of Van Hise Hall, a delightful rock garden is laid out around the trunk of an impressive bur oak. The tree, called the Euthenics Oak, is actually a replacement for the original white oak, native to the site, which died.

Abby L. Marlatt, professor of home economics from 1909 to 1939, and director of the school for a quarter of a century, established the garden in 1925 with the assistance of her students. At that time they named the tree the Euthenics Oak in honor of the society of the same name, a home economics organization dedicated to the improvement of human life through adjustment of the environment.

The American Yellowwood

Madison

One of the rarest trees in the state is the American yellowwood. The second largest of its kind in Wisconsin grew in the courtyard of the Adams Residence Hall on the University of Wisconsin–Madison campus until it came down in 1999 due to structural failures. Its circumference was 5 feet 8 inches. A southern Appalachian native, it covers itself with beautiful white blossoms every spring. Three other yellowwoods existed previously on the campus, but the somewhat protected location of the Adams Hall specimen apparently insulated it from winter blasts, allowing it to survive longer than the others.

The Edgewood Oaks

Madison

Until the early 1990s, a venerable bur oak leaned out from a fringe of woods, arching its strong, heavy limbs across Edgewood Drive, the tree-shaded lane that leads from Woodrow Street into Madison's Vilas Park. It was one of the oldest trees in the capital city, with a circumference of 10 feet 10 inches. Early Madison historian A. O. Barton estimated the tree's age at 400 years. Walter E. Scott, noted big tree recorder, says Increase Lapham, Wisconsin's first scientist, may have tethered his horse Adelaide under this tree. Certainly Lapham spent time on that very ridge, surveying the Indian mounds and preparing a map of them to be published by the Smithsonian Institution in *Antiquities of Wisconsin* in 1855.

According to Prof. Robert S. Ellarson of the University of Wisconsin, the most widespread and abundant plant associations the first European explorers and surveyors found in the Four Lakes area were oak openings with the bur, white and sometimes black oak the dominant members. Other surveyors, working during December 1834 where the state capitol now stands, wrote in their field notes: "timber, black, white and burr O. Undergrowth same."

In 1976, Walter Scott, whose interest in trees had raised everyone's natural heritage consciousness, proposed as a Bicentennial project to identify all the 200-year-old oaks within an eight-mile radius of the capitol. A total of more than 400 oaks met or exceeded this size requirement and were recorded and marked. Seven of them, including the Lapham tree, were on the Edgewood campus.

As the historian Barton wrote of the oldest Edgewood oak in the 1930s, "It has seen the coming and going of waterfowl on Lake Wingra, sheltered generations of Indians, and watched the deer and buffalo feeding in the present Edgewood pasture. It was an old tree when Madison was settled a century ago."

All the Edgewood campus oaks presided over ages of Native American encampments, as the Ho-Chunk made their annual migrations to Lake Wingra, one of their favorite hunting and fishing grounds. While the Ho-Chunk presence diminished after the Black Hawk War of 1832, small numbers continued to come back each spring, well into the 1900s, their temporary summer shelters suddenly appearing along the shoreline.

In the 1840s pioneer governor Leonard Farwell bought up hundreds of acres of the wooded lands around the lake. In 1855 he sold 55 acres of the higher ground to John Ashmead, a Philadelphia lawyer. Ashmead called the beautiful villa he built there for his family "Edgewood." When his wife died only a year later, he sold the place to Samuel Marshall, who had founded the Marshall and Ilsley Bank of Milwaukee and wanted to start a bank in Madison. Marshall, an amateur horticulturist, planted the stately row of trees on each side of the driveway from Monroe Street to the house. In 1873 Governor-elect Cadwallader Washburn became owner of the estate. Unfortunately he failed to win a second

Edgewood Drive Oaks B-Wolfgang Hoffmann (1982)

Edgewood Drive Shagbark Hickory B-Wolfgang Hoffmann (1982)

term, and he decided to move to Minneapolis. In a most generous gesture, he gave the property to the Dominican Sisters of Sinsinawa, who at that time operated St. Regina's Academy on Henry Street. After the details of the turnover had been settled, Governor Washburn, proud of the landscaped grounds, escorted the sisters on a tour, pointing out the many beautiful trees: a shell bark hickory, many pines and a huge silver maple. He asked that they be saved and protected. Of course they were. Many of the lovely and historic old trees continue to add immeasurably to the tranquil learning environment of Edgewood's modern-day educational complex.

Sources: Claire Geesaman, Madison
Kit Hildebrand, Madison

University of Wisconsin Arboretum

Madison

Establishing an arboretum can be a complicated and demanding undertaking. The needed interplay of skills and understandings, expertise and common sense, personality and perseverance is as rare as some of the vanishing species and plant communities an arboretum is formed to protect.

The names of the gardens, prairies, drives, woods and individual trees in the University of Wisconsin Arboretum in Madison are a litany of the people who made it happen. From the first mention in 1909 by landscape architect John Nolan, recently arrived from Massachusetts, and the first specific proposal placed before the Board of Regents by Michael B. Olbrich in the mid-1920s, to the present day, numerous practical dreamers have overcome incredible obstacles to ensure that Wisconsinites enjoy a delightfully restorative and valuable resource which will be preserved for future generations.

Well before Nolan and Olbrich, Increase Lapham, the state's first scientist, foresaw the need for preserving such a natural community in 1853 when he wrote in "The Forest Trees of Wisconsin" that "It would seem to be peculiarly appropriate for our Universities and Colleges to secure upon the grounds by which they are surrounded, at least one good specimen of each tree and shrub that grows naturally in Wisconsin; and I will venture to predict that the University or College that shall first surround itself with such an 'Arboretum' will first secure the patronage and good opinion of the people, and will thus outstrip those institutions that show a lack of taste and refinement by omitting to plant trees."

Equally important to planting trees is the need to study living plant communities, their interactions and evolution.

Even after the idea of an Arboretum was well accepted, the acquisition of the land, over 30 or more years (it is still going on), and the raising of needed funds were hectic and fraught with setbacks. But the diverse terrain now comprising the 1,200 acres holds perhaps more potential scientific opportunity than careful planning could have accomplished. "No other experiment in the country contains the number of different plant communities as does the University of Wisconsin Arboretum," according to Nancy D. Sachse, author of *A Thousand Ages.*

Today a foot traveler through the Arboretum can revel in a variety of natural environments: diverse types of wetlands (too fragile for casual visitors), shrub, savanna, conifer swamp, boreal forest, grassland, prairie, deciduous forest, pine forest, horticultural areas and experimental plots. Birders and botanists, wildlife students and ecology buffs have here an outdoor laboratory, workshop and display area. In the spring, for instance, people come from all over the Midwest to see and breathe in the heavenly fragrance of blooming lilacs and flowering crab apple trees in the Longenecker Gardens, named for the man who was the executive director of the Arboretum from its founding in 1932 until his retirement in 1965.

But the Arboretum is also the domain of scientists and students. Dozens of experiments of all types are conducted, while opportunities for continuous observation, equally important to naturalists, are many. As one example, Noe Woods, named for the donors of the land, can answer some vital questions about the life of an oak forest: What will happen when the dominant oaks, now mature, begin to age and die? Will the oaks predominate through only one generation? What species of tree—red maple, young white oak, black cherry—might succeed them? Since no

Jackson Oak at the University of Wisconsin Arboretum

B-Wolfgang Hoffmann (1982)

one has ever lived with an oak forest long enough to find out what becomes of it in the long run, the Arboretum, protected from disturbance and available to succeeding generations of researchers, can help shed light on these questions.

One of the treasures of the Arboretum was the Jackson Oak, a great, spreading tree, typical of oaks grown in full-sun openings. It was a seedling about the time Madison was chosen as the state capital in 1837, when at least 500 Native Americans were still camped around the Four Lakes. According to the late University President E. B. Fred, writing in *A University Remembers,* the oak "resembles in character the man it is named after, Colonel Joseph W. Jackson, an Arboretum founder and dedicated and untiring crusader for Arboretum fund raising and land acquisition." Sadly, the Jackson Oak, whose gnarly arms raised above the Curtis Prairie became a symbol of the Arboretum, died in the late 1990s.

The Leopold Pines, a forest of red, white and jack pines planted during the late 1940s and early '50s are, of course, named for Aldo Leopold. This world authority on game management and ecology was a member of the first Arboretum Committee and gave unstintingly of his time, efforts and expertise in restoring and establishing the resource. To him, an arboretum meant "far more than a mere collection of trees," but rather "a sample of what Dane County looked like when our ancestors arrived here." Recently, Arboretum officials have been clear-cutting some of the crowded pines and planting red maple, red oak, aspen, paper birch and basswood to create more of the diversity typical of native pine forests.

The Gallistel Woods, honoring Albert Gallistel, a charter member and 20-year chairman of the Arboretum Committee, include several kinds of deciduous forest. The Wingra Woods include the Arboretum's most magnificent stands of oaks, its preponderance of red oaks distinguishing it from Noe Woods. On its north-facing slope, Arboretum developers have planted hemlock and yellow birch, sugar maple, basswood and beech, all species characteristic of northern mesic forests, which were, at the time of settlement, Wisconsin's largest single plant community.

Sources: William R. Jordan, II, Madison
Jim Zimmerman, Madison

Whitnall Park and Boerner Gardens

Milwaukee

Milwaukee County's Whitnall Park and Boerner Botanical Gardens are a testament to the philosophies and foresight of two men dedicated to the concept that natural beauty should be available to every city dweller. As a result, many varieties of handsome and important trees have been protected, nurtured and planted in the Hales Corners park and adjoining Root River Parkway system, a total of 3,266 acres inside Milwaukee's city limits.

One of the state's oldest sugar maples—over 270 years old—is growing near the corner of Grange Avenue and Root River Parkway Drive. Many other oaks and maples 100 to 200 years old are likewise still thriving, along with mature ashes, basswoods, shagbark and bitternut hickories, black walnuts and hawthornes. The stately American elms, regrettably, have mostly fallen to the infamous Dutch elm disease.

According to Charles B. Whitnall, the "father of the Milwaukee County Park system," who died at the age of 90 in 1949, a rural park should be very large and developed so that as much of it as possible can be left in a natural or semi-natural state. These extensive park and parkway areas provide not only recreation, but educational and inspirational advantages for a fast-growing community. A man well ahead of his time, he said, "We are badly in need of social adjustment, in which 'Land Use' appears to be a most vital factor, because with the natural landscape undefiled, the physical influences on which we depend preserved, a desirable environment is possible."

Alfred L. Boerner, a landscape architect, also believed that a park should provide esthetic gratification and recreational and educational opportunities. With great effectiveness, he combined beautiful formal gardens with informal areas stressing the natural beauty of the Wisconsin terrain.

Ironically, the Depression made it possible to bring to fruition the combined philosophies of the two men. Civilian Conservation Corps, Works Projects Administration, National Youth Administration and county relief workers accomplished the building of nature and hiking trails, roads, parking lots, structures, dams and bridges, as well as the landscaping and tree planting, during the years 1932 to 1941.

Boerner conceived his Botanical Gardens as, first in importance, a collection of trees, shrubs and other woody plants, with the formal gardens as secondary. The plan was to create a museum of woody plants, both native and exotic, arranged according to plant families, for display, research and nature study. Not the least of his educational aims was to enable citizens of Milwaukee County to learn from example how to make their own home sites more attractive. Thus, unusual trees from all over the world would be tested for aesthetic appeal and hardiness in this climate.

Various groups undertook a number of individual memorial tree plantings, marked by plaques, during the 1930s. These included an ancestral grove established by the Milwaukee County Genealogical Society. Among other outstanding sights in the Boerner Gardens are the Scots pines, planted about 100 years ago, that formed the nucleus of the present expanded conifer collection on 92nd Street near College Avenue; the flowering crab apples and lilacs; a collection of nut trees started in 1939; a collection of dwarf fruit trees started in 1957 and presently containing 29 varieties of apple, cherry, pear, peach and plum; an experimental street tree planting, accomplished from 1967 to '69, that follows new ideas in selection, species and varieties; and the Potter Forest, located on the northern boundary of the park, that is enjoyed by Milwaukee school children as a nature study area.

Source: Mrs. Marshall Thompson, Milwaukee

The Shorewood Tree Collection

At least 11 champion trees and many unusual ones made up the collection planted on the grounds of the former Shorewood Sanitarium at 2316 E. Edgewood, in the Village of Shorewood in Milwaukee.

There are two ancient Indian trails through the property. The Daughters of the American Revolution marked one of them, a Sauk Indian trail that followed along the Milwaukee River, a part of the early route from Chicago to Green Bay.

William Studley, medical director of the sanitarium from the early 1930s until the institution closed in 1977, oversaw the planting of most of the trees. Studley's father Frank founded the sanitarium in a building across the river from the present site in 1898. In 1904, workers began construction of the new building, which eventually came to be owned by Columbia Hospital.

Among the champion trees on the grounds during the 1980s were the state's largest striped maple, persimmon, sour cherry, Japanese zelkova and London planetree. Second rank in size went to a sycamore, maple, pecan, eastern wahoo and the magnificent dawnredwood. The last, *Metasequoia glyptostroboides,* was reclassified as to its genera in 1941 when Shigeru Miki, a Japanese paleobotanist, noted the opposite pattern of its needles. The trees were found growing in China during World War II, and afterwards some of them were exported to the United States. In the category of third largest, a sweetgum took the honors.

In addition to these record trees, a paperbark maple, a yellowwood, a bald cypress, and numerous other exotics as well as many large, old native trees graced the hospital grounds. The grounds were sold and subdivided in the 1990s. At the urging of many concerned citizens, the village required that as many of the trees be preserved as possible. However, development and change have taken their inevitable toll, leaving just a remnant of the Shorewood collection.

Source: William Studley, Milwaukee

Whitewater College Arboretum

Wisconsin Historical Society, WHi(X3)12686 (c. 1930)

The Whitewater College Arboretum

The University of Wisconsin–Whitewater campus boasted three trees that were state records during the 1980s: a Pekin lilac, a cucumbertree magnolia, and a shingle oak. Professor Salisbury, who became president of the State Normal School for the preparation of teachers in 1855, conceived the idea of "making the school grounds serve the purpose of an arboretum, establishing therein every species of tree or shrub that will endure this climate."

By October 1908, more than 150 species were already to be found on the school premises. The planting had begun on May 1, 1869, a year after the school's founding, when the male students assembled and planted a row of elms along Grand Avenue, "each man planting one or two trees, the cost of which he bore. At the close of this exercise, a series of wrestling matches was indulged in by the young men, and a spread was served by the young ladies of the school."

In addition to the former state record-holding trees, the campus includes large and handsome specimens of sycamore, buckeye, ginkgo, southern catalpa, swamp white oak, pin oak, and many others, including a Schwedler's maple planted by Professor Salisbury's own hand.

The front campus is called the Chopp Arboretum. The north and northeast areas include Friar's Woods and the Drumlin Arboretum.

Source: Maurice Kalb, Whitewater

The Trees at Yerkes Observatory

Williams Bay

At one time in the 1980s, there were at least seven state record trees on the splendid park-like grounds of Yerkes Observatory on the shores of Lake Geneva in Williams Bay. This imposing site is still the resident observing facility of the Department of Astronomy and Astrophysics at the University of Chicago.

It was the dream of George Ellery Hale, who owned a small private observatory, to establish a larger one accessible from Chicago, but far enough away to benefit from clear night skies. With the active cooperation of William Rainey Harper, the first president of the University of Chicago, Hale prepared the plans and enlisted the financial backing of Charles T. Yerkes, a wealthy Chicago businessman. Construction in Williams Bay began in 1895, and when finished, the observatory was the most modern and complete one

Trees at Yerkes Observatory Wisconsin Historical Society, WHi(X3)38159 (1935)

of its day. Still important, its great refractor telescope is the largest one in the United States.

During the period from 1912 to 1915, the landscape architect firm of Frederick Law Olmsted drew up a landscaping plan for the grounds. Olmsted, pioneer landscape architect in the nation, was famous for his design of Central Park in New York City. The Yerkes plan was a grandiose one, and its implementation was somewhat scaled down. But Edwin Frost, director of the Observatory at that time, who loved trees almost as much as he loved stars, was enthusiastic about carrying it out.

Besides the state record trees, which were a white fir, yellowwood, goldenraintree, European and fernleaf beech, Japanese pagodatree and a European littleleaf linden, the grounds boasted other interesting trees such as Kentucky coffee trees, ginkgos, cut-leaf beeches and chestnuts.

Sources: Judy Lola, Williams Bay
Dr. W. W. Morgan, Williams Bay

The Mount Vernon Forest of Fame

Dane County

There is a Forest of Fame in Mount Vernon in Dane County near Highway 92. Appropriately enough, a group of elm seedlings brought from George Washington's Mount Vernon in Virginia (for which the Dane County settlement was named) were the first trees planted there to celebrate Arbor Day in 1916. The people of the neighborhood gathered for the ceremony and listened to an address marking the occasion by the Hon. John Sweet Donald, then secretary of state for Wisconsin, who started the forest.

Donald, called a "friend to man and nature," was also a professor of agricultural economics at the University of Wisconsin in Madison. His dream was to restore a field he had bought in the small village to the beautiful forest it had once been. Over the years he planted trees from the birthplaces of former presidents of the United States, famous generals of the world, and personages associated with religion, science, music, agriculture and commerce. Inscrip-

Mount Vernon Forest of Fame

Wisconsin Historical Society, WHi(X3)38167 (c.1933)

tions were painted on boards mounted in concrete near each tree.

Among the trees, besides those from Washington's home, were elms planted to honor Robert La Follette, Woodrow Wilson, and Wisconsin governors W. D. Hoard, Lucious Fairchild, Cadwallader Washburn and Leonard J. Farwell; an elm from Abraham Lincoln's farm in Hodgeville, Kentucky; and the Liberty Tree, from Blue Mounds, Wisconsin, in memory of Wilhelm Tell.

The MacKenzie Environmental Education Center

Poynette

What used to be called the State Game Farm at Poynette, and is now named the MacKenzie Environmental Education Center, offers visitors many riches, including an arboretum planted with more than 200 species of trees, some of them of state record size.

The state has set aside over 500 acres here in Columbia County, under Department of Natural Resources jurisdiction, for study, recreation and conservation work in hopes of improving people's relationship with nature. Facilities, staff and equipment are available for intensive programs in observation and study of the environment by elementary, secondary and college students and environmental groups.

The educational potential of the state game farm concept was recognized early and led to the development of demonstration conifer plantings and the arboretum. Among the record-sized trees that can be viewed there are three varieties of pine, three varieties of oak, an amur maple, silverbell, bigleaf linden and Black Hills spruce. Unfortunately, a state record osage orange had to be removed recently.

Star Lake Forest

Vilas County

In 1913 in Vilas County, E. M. Griffith, Wisconsin's first state forester, guided the establishment of the Star Lake Forest Plantation, another Wisconsin first. Griffith considered the plot, which jutted out into Star Lake, an experimental plantation to be used as a guide for future tree planters in Northern Wisconsin.

Griffith was indeed ahead of his time. In 1915, the State of Wisconsin declared forestry unconstitutional and abolished Griffith's position as state forester. Today, records from the plantation provide data on tree growth and thinning operations.

Sources: M. N. Taylor, Merrill
Paul Pingrey, DNR, Madison

Star Lake Forest Plantation (c. 1946)

Trees for Tomorrow

Vilas County

Ever since bearded explorers tramped the forest wilderness of Wisconsin, trees have played a functional role in the economic, biological and social life of this state. A unique organization, Trees for Tomorrow, has played a part in shaping that role.

Trees for Tomorrow was founded on a wintry day, February 29, 1944, in Wausau. M. N. Taylor, advertising manager of the *Merrill Daily Herald,* was executive director. The other original officers were paper company executives Folke Becker, Rhinelander Paper Company, president; Sven Bugge, Tomahawk, Kraft Paper Company, first vice president; and David B. Smith, Merrill, Ward Paper Company, secretary and treasurer.

At that time, 85 percent of wood consumed by Wisconsin paper mills was imported from other states and Canada. Wisconsin's landscape had been stripped of its virgin pine cover. Remnants of less desirable species, containing fiber for gunpowder and thousands of other uses, could have been used to help win the war. On the other hand, vast fields of barren, non-agricultural land lay idle. Cut-over acreage was tax delinquent or selling for $1 an acre.

It was in this wartime climate, the fall of 1943, that the Wisconsin wood-using industry sparked a crusadelike campaign to stimulate the production of pulpwood for the war. Some mills, due partly to lack of manpower, were down to a week's supply of wood.

The *Merrill Daily Herald* dramatized what could be accomplished through coordinated effort by organizing a Pulpwood Roundup and publishing a Pulpwood Tabloid. The War Production Board in Washington, the Wisconsin wood-using industry, Forest Service, State Conservation Department and local agencies jointly sponsored this effort to stimulate the production of pulpwood. Women cut pulpwood, Sunday sermons in church stressed the shortage of wood, county agents contacted farm woodlot owners, war bonds were offered as prizes for best-appearing and largest loads of wood at the Roundup. When 125 trucks of pulpwood rolled down the main street of Tomahawk, paper industry executives reasoned that if this pent-up energy could be channeled into the planting of trees, Wisconsin might again be covered with new forests. The result was that M. N. Taylor, who had spearheaded the pulpwood campaign, was named head of a committee to implement a long-range Trees for Tomorrow program.

The first step was to become known and accepted. Initial projects included: a two-day Reforestation Institute at Rhinelander; hand planting of 10,000 trees

Trees for Tomorrow Camp: W. A. Sylvester, William Bjorge and M. N. Taylor (c. 1948)

by high school students; the organization of machine planting; and the distribution of free trees with 200,000 seedlings handed over to private landowners.

In 1945, the 2-for-1 program was introduced. Two trees were given free for each tree planted. Landowners, schools and communities planted four million trees in the Wisconsin River valley. Forty-two school and memorial forests were dedicated. As many as 730,000 free trees were distributed each spring. There were no strings attached. Trees could be planted for any forestry purpose.

Before long, landowners sought help in managing their second growth forests. The first of five foresters, William Sylvester, had already joined the staff. Three foresters were now employed in Merrill and two in Eagle River. At first, the organization offered forest management services free of charge. Strangely enough, when a small fee was charged, requests for management services increased 56 percent. The fee put a value on the work.

Trees for Tomorrow became a catalyst for action programs, which provided an opportunity for the involvement of others. An example: the Wisconsin Bankers Association sponsored forestry field days. They bought 24 tree-planting machines in the Wisconsin River valley and planted millions of trees.

Landowners began to ask for help with timber sales. Trees for Tomorrow added assistance in harvesting timber to its services. Foresters cruised a piece of property, prepared a sales prospectus, contacted producers, asked for bids, secured a performance bond and made sure the operation was covered by insurance. Again, they charged a small fee.

Many changes were taking place in wood procurement. Sophisticated machinery was going into action, but there was a shortage of woods workers. So, Trees for Tomorrow set up the first all-around loggers' apprenticeship training program in the Lake States at Eagle River and initiated a loggers' course for Nicolet College at Rhinelander.

During the span of the forestry program, Trees for Tomorrow had distributed and machine-planted 23 million trees; prepared management plans for 370,000 acres of private woodlands; and estimated, marked and supervised the sale of 440,000 cords of wood.

But the time arrived for Trees for Tomorrow to evaluate its forest management and tree-planting program. The plain fact was that Northern Wisconsin was running out of open fields on which to plant trees. More field services were also becoming available from public agencies. So, a chapter in the history of forestry in Northern Wisconsin came to an end. The Trees for Tomorrow Camp, founded in 1946 at Eagle River, however, still functions as a resource education headquarters. And the idea of self-help, still sponsored by industry and a wide cross-section of the public, continues to be a viable one.

Source: From an article by M. N. Taylor, Merrill

The Paine Arboretum

Oshkosh

When Nathan Paine, grandson of a pioneer Oshkosh lumberman, began to build his English Tudor home in 1927, he tended to every detail. He was so concerned that construction might damage the roots of a great bur oak in the back of the mansion that he personally saw to the redesign and relocation of the service entry. Until it blew down in 1996, this bur oak was among the outstanding features of the arboretum and gardens that contributed to the estate's tranquil country setting within the city.

Paine built the mansion to house himself and his wife as well as their extensive and important art collection. He planned that on their deaths it would be donated to the city of Oshkosh as an arts and cultural center. Due to delays in construction caused by the Depression and World War II, however, neither he nor his wife ever lived there. After Paine's death in 1947, Jessie Paine oversaw its completion and establishment as one of the first 200 museums in the country to be accredited by the American Association of Museums. In 1978, 30 years after it opened to the public, the Paine Art Center and Arboretum was placed on the National Register of Historic Places as one of the finest examples in the country of an English manor house and grounds. It is now known as the Paine Art Center and Gardens, reflecting the loss of many of the larger trees, but still emphasizing the horticultural value of the many gardens.

New Yorker Bryant Fleming, the first instrument of Nathan Paine's dream, was both architect of the home and the landscape architect. Together, they studied the environment and the demands of the climate and chose a variety of woody ornamentals for their springtime blossoms, fruit-laden boughs or brilliant autumn foliage to be planted on the 14-acre complex. They considered color and texture and shape in achieving a harmonious arrangement. Mark Hopper was the first fulltime manager of the arboretum, followed by John Green. Some of the species still adding such beauty to the grounds are red jade weeping crab, columnar white pine, dawnredwood, European beech, pagoda dogwood, Douglasfir, ginkgo, European bird cherry, Japanese tree lilac, saucer magnolia and Koster blue spruce.

Green Gables Estate

Lake Geneva

The crystal-clear waters and wooded shores of Lake Geneva provided the ideal Wisconsin retreat for many of the wealthy nineteenth century Chicago families. William Wrigley chose a lush shorefront site along Snake Road to build his summer mansion in 1863. It took two years to complete, using an elaborate architectural design and stone construction. The gatehouse was most remarkable for its steep roof gables made of green slate. The estate was quickly dubbed Green Gables. At one time Green Gables was home to three Wisconsin Champion trees—a catalpa, Douglasfir and boxelder.

Longtime groundskeeper, the German-born Wally Koehn, now deceased, reported that the boxelder, which had been held together with tons of concrete and many feet of cable, finally succumbed and was removed by the Rudy Lange Tree Service in 1980. The majestic Douglasfir and catalpa, growing in the front lawn and once visible from the lake, are now gone.

Wally Koehn beside catalpa at Green Gables Estate

B-Wolfgang Hoffmann (1979)

Chapter IX

Trees in Literature, Art and Folklore

Ben Logan's Big Maple Tree B-Wolfgang Hoffmann (1982)

Ben Logan's Big Maple Tree

Crawford County

Ben Logan, well-known author of *The Land Remembers,* a memoir of his boyhood on a Wisconsin farm in Southwestern Wisconsin, wrote compellingly about a big soft maple tree that stood on a ridge near the house. The whole family plus the hired man opposed Logan's father's plan to cut the tree down. A caring but logical man, he feared a storm would send a branch crashing through the roof. The following passage, quoted with permission from Viking Penguin, Inc., described the conflict:

The tree was important to each of us in different ways, but most important to Mother. I think she saw it as a place where we lived, an extension of the house, and the house was her domain. The maple tree was the first sign of the farm you could see as you came over the top of the hill leading up from Halls Branch Valley. When we'd been away on Sunday and there had been a windstorm, she would lean forward as the Model T groaned up the last steep pitch to the ridge. Then she'd lean back, smile, and say, "It's still there."

Most of all, the tree linked us to the migrating birds and the changing seasons. The new birds came to its leafless branches when its blossoms were a fringed silhouette of red and yellow, a week or more ahead of any other flowers.

The tree was more to her than we knew. Many times a day, looking up to where its high branches caught the breeze, she began to sing. There was something very warm and right about that, reminding us that Mother, too, had her own private world.

I'm not sure Father realized how important the tree was to Mother. If he had, I don't think he would have kept wanting to cut it down. It wasn't that Father didn't like the tree. He did, and he enjoyed the good times we had under it; but he didn't trust the tree.

"Look," he would say, pointing at the heavy limbs, "soft maple trees don't belong on the ridge in the first place. They can't take the wind storms."...

Finally it happened. There was a storm. A limb split loose from the trunk. There was a crash. The house shook. Again, the limb had missed the chimney, but for Father the fallen limb was proof that the tree should go.

After breakfast, out came the crosscut saw. He sharpened it carefully, decided which direction the tree should fall, and called Lyle and Junior. He held out the saw....

Mother stood on the porch, face set, arms folded. She said nothing....

Lyle and Junior knelt beside the tree to start the first cut.... Father stood over them.

Mother was still on the porch. I couldn't understand her silence.

The saw cut deeper into the trunk. It had eaten in about five inches when Lyle and

Junior looked at each other and came to some agreement.

"Time for a rest," Junior said.

Father moved closer. "What's the matter?"

"We're tired," Junior said. His voice was polite but very firm.

Lyle looked around and met Mother's eyes. For just a minute his crooked grin was on his face. "As a matter of fact," he said, "we're damn tired. It just might be about fifty years before I'm rested up enough to try that again."

Mother smiled.

Father looked from Lyle to Junior to the tree. We waited knowing how much he counted on being obeyed. For the first time he looked at Mother. She tried to hide her smile. She didn't quite make it,

"Look," Father said. "That damned tree... I don't..."

"I know," she said.

Father took a deep breath, and slowly let it out again. He took the saw out of Lyle's hands. We still weren't sure. One man could handle a good crosscut saw if he had to.

Tooth by tooth, Father brushed the white sawdust from the saw. He walked to the milkhouse and hung it up in its usual place....

Only once after that do I remember any mention of the tree. The morning after a bad storm, Lyle looked into the yard and found no limbs on the ground. He forgot himself. "Hell, that tree's going to outlive us all."

"Maybe because it knows how we feel," Mother said, smiling.

Father looked at her for a long moment. He smiled. "Could be that. Could be I scared it a little, too."

The tree is still there, four feet through now, the biggest soft maple I've ever seen on the ridge. The scar, low to the ground on the south side, is hardly noticeable.

Logan's resonant prose stands by itself, expressing the deeply intuitive appreciation of trees that has characterized many of the people whose stories are in this book. An interesting postscript involves a nearby family's love of trees that went back even earlier.

After the publication of *The Land Remembers* in 1975, Logan got a letter from an old neighbor, Josephine Mullaney, who had been born in the one-room log house across the draw from the Logan farm in 1886. In a letter rich with the recollection of bygone years, she recalled her home: "A piece of me is most surely left on that ridge farm in Haney. Could be in a clearing in the white oak woods just south of the bridge, where the bird's foot violets bloomed."

She went on: "And I could well have been the one who planted the soft maple tree you loved. We all loved trees—the white pines were likely brought from our Uncle's, John Lynch, on Copper Creek or by Petersburg. It is unusual, I believe, for a soft maple to be so long-lived and to attain the size you indicate."

Not so unusual, perhaps, for a tree that is loved.

Source: Ben Logan

Aldo Leopold's Good Oak

Sauk County

Aldo Leopold is now recognized as one of the country's pioneer conservationists, the father of the profession of wildlife management, and the first to poetically enunciate the land ethic. Leopold was a professor of game management at the University of Wisconsin–Madison when he acquired an 80-acre farm in Sauk County in the late 1930s. The Shack, the only building on the place, sheltered the Leopold family on their weekend trips to the farm to undertake one of the classic adventures in ecological restoration.

In *A Sand County Almanac,* his beloved account of bringing this corned-out, abandoned, sandy-soiled failure to an ecological balance, a place of beauty, a laboratory for study, and a "weekend refuge from too much modernity" (as Leopold biographer Susan Flader describes it), Leopold told of planting and living off the land. His description of felling an old oak tree for firewood is one of the most memorable passages in Wisconsin (or any) literature:

It was a bolt of lightning that put an end to woodmaking by this particular oak. We were all awakened, one night in July, by the thunderous crash; we realized that the bolt must have hit near by, but, since it had not hit us, we all went back to sleep. Man brings all things to the test of himself, and this is notably true of lightning.

Next morning, as we strolled over the sandhill rejoicing with the coneflowers and the prairie clovers over their fresh accession of rain, we came upon a great slab of bark freshly torn from the trunk of the roadside oak. The trunk showed a long spiral scar of barkless sapwood, a foot wide and not yet yellowed by the sun. By the next day the leaves had wilted, and we knew that the lightning had bequeathed to us three cords of prospective fuel wood.

We mourned the loss of the old tree, but knew that a dozen of its progeny standing straight and stalwart on the sands had already taken over its job of wood-making.

We let the dead veteran season for a year in the sun it could no longer use, and then on a crisp winter's day we laid a newly filed saw to its bastioned base. Fragrant little chips of history spewed from the saw cut, and accumulated on the snow before each kneeling sawyer. We sensed that these two piles of sawdust were something more than wood: that they were the integrated transect of a century; that our saw was biting its way, stroke by stroke, decade by decade, into the chronology of a lifetime, written in concentric annual rings of good oak.

Source: Robert McCabe, Madison

The Most Painted Tree in Wisconsin

Milwaukee County

During the time when loggers were cutting trees in Wisconsin by the thousands, one tree was becoming famous as the most painted tree in the state. It was a willow on the farm near Milwaukee of German-born artist Henry Vianden.

A man who loved nature, Vianden farmed to cut expenses, while making most of his income from painting. He had revealed his artistic talent as a young child, had apprenticed to a goldsmith, and later worked as an engraver and a painter on porcelain.

In America, Vianden instructed students in art, holding classes outdoors near his cottage. By one account he once offered this criticism to a budding talent: "That? That should be a tree? A tree, that is

Aldo Leopold

Robert A. McCabe (1946)

the most wonderful thing in the world. When we paint a tree we must realize that we have sitting for us, a king!"

In his own painting of trees, Vianden's technique was to render every leaf and every branch with precise exactitude. He wanted his subjects to be admired in all their detail. An excellent example of his work, "The Old Oak," was once owned by the *Milwaukee Journal.* Today its whereabouts are unknown.

Vianden died in 1899. He not only immortalized the much-painted willow tree, but gave countless students an appreciation of art and of the masterpiece that is a tree.

Source: Steve Stroud, Milwaukee

Caddie Woodlawn and the Old Elm

Dunn County

A huge old elm in the village of Dunnville in Dunn County stood in the yard of "Dr. Crocker's house." In 1900, reports Christine Granger Klatt, who was there, it was the largest tree in the village. Its associations are both literary and historic. The significance of Caddie Woodlawn Park has been documented by an official Wisconsin Historical Society marker.

Dr. Walter Crocker, the first doctor in the area, treated patients as far away as Chippewa Falls and Menomonie and all through the pineries and logging camps. He built a home for himself and his family in Dunnville in 1856, when the dooryard elm was planted.

Oddly enough, Dunnville today is perhaps better known because of its literary connections than its historic importance. In the early 1850s, the village, seemingly, had everything going for it. As the county seat, it was ideally situated at the juncture of the Menomonie and Red Cedar Rivers and both rail and steamboat service were to link it with the rest of the state. But a fire in 1858 burned down the new wooden courthouse with all its records. Two years later Menomonie became the county seat and Dunnville settled into slow decline.

Fortunately, the fire spared Dr. Crocker's house and the elm tree. In 1900, 11-year-old Christine Granger moved with her family to Dunnville. Mrs. Klatt recalled playing under the huge elm. By that time the Crockers had relocated to a farm on the Dunnville Prairie, destined to become even more famous in literary history than the village itself.

Christine, her sisters and friends played make-believe housekeeping games under the stately elm and had tea parties, with leaves and seeds for food and water for tea.

How did Dr. Crocker, planter of the elm tree, achieve literary fame? Through two books written by Carol Ryrie Brink. A description of Mrs. Crocker's doll collection appears in *Magic Melons,* a story for young people. Mrs. Brink is much better known, however, for her Newbery Award-winning book, *Caddie Woodlawn,* written in 1935 and based on her grandmother Caddie Woodhouse's stories of her pioneer Wisconsin girlhood on the Dunnville Prairie. The Crocker farm adjoined the Woodhouse (Woodlawn in the book) farm.

Mrs. Klatt and her sisters even attended the same school Caddie and her brothers and sisters had gone to between 1860 and 1867.

Source: Christine Granger Klatt, Menomonie

The Durward's Glen Spruce

Sauk County

A Scottish-born poet and painter, Bernard Isaac Durward, came to Wisconsin in the middle 1800s and settled himself in 1862 in a lovely, isolated valley in the Baraboo Hills, Sauk County, now called Durward's Glen.

In 1852 he painted a portrait of Bishop Henni and became a Roman Catholic. His home reflected a life dedicated to God, nature and art. Among his plantings was a Norway spruce, which grew to be the third-largest in the state, third only because its trunk forked before the official circumference-measuring height of 4½ feet. At 3 feet, its girth was 13 feet 8 inches. Apparently the tree is now dead.

Durward's paintings and those of his equally talented son, Charles, hang in the Rock Gallery at the Glen. The estate is now the home of the Saint Camillus Novitiate.

The Father Hennepin Blaze Tree

Racine County

Increase A. Lapham and Philo Hoy, M.D., were among the earliest scientists and natural historians of Wisconsin. In the late 1860s, Dr. Hoy heard that a large oak tree had blown down on a farm near Racine. He and Lapham, who happened to be in the vicinity, went to the farm and watched as the farmer sawed the tree into usable lengths.

In the middle of the tree, not too far above the ground, the two scientists noted a curious phenomenon: what appeared to be the mark of an ax inside the tree, as though at some time someone had used it to blaze a trail. Hoy and Lapham counted the rings from the surface of the tree to the blaze. They conjectured that the blaze had been made during the years in which the Jesuit explorer Father Hennepin made his journey up the Mississippi. It is not recorded what happened to the strange blaze mark, but it probably burned up in the farmer's stove.

Prairie du Chien's Black Hawk Tree

Anyone familiar with Prairie du Chien knows the most famous local tree there was the old cottonwood called the Black Hawk Tree. Generations of residents have believed the legend that Black Hawk, the renegade Sauk Indian leader, once hid in this tree—perhaps sometime in the 1790s—to evade capture by soldiers of the Fort Crawford military post.

For many years the "middle-of-the-road" tree served as a unique and popular landmark. Even after a windstorm took it down almost 75 years ago, a marker showing its location was imbedded in the surface of the highway where it had stood. That street is now the main east-west business street of Prairie du Chien and is named, appropriately enough, Black Hawk Avenue.

Many historians believe Black Hawk was in Prairie du Chien only once: after his surrender at Wisconsin Dells to the Ho-Chunk, following the Battle of Bad Axe, 40 miles north of Prairie du Chien, which ended the 1832 Black Hawk War. He could not have hidden in the tree then because he was already in custody. He was held briefly at Fort Crawford by General Street, and then a detail headed by Jefferson Davis took him, encumbered by ball and chain, to St. Louis. His imprisonment was short, and he made a triumphal tour of eastern cities, lionized by ladies, and became in fact, in the words of Anthony F. C. Wallace in his introduction to *The Black Hawk War,* published by the Illinois State Historical Library, "the prototype of the noble redman, defending the honor of a dying race."

At any rate, whether the Black Hawk Tree served as Black Hawk's hiding place or not, it was a rarity in a town that had both few trees and a sizable population using wood for fuel. There is no question that the Black Hawk Tree was the best-known tree in the area, a part of the lore of the town, and as such, is deserving of recognition.

Sources: Jack Holzhueter, Madison
Crawford Thayer, Fort Atkinson
Griff Williams, Prairie du Chien

Prairie du Chien's Black Hawk Tree

Wisconsin Historical Society, WHi(X313)2623 (c. 1905)

Scott's Mansion F. Mead (c. 1900)

Scott's Mansion Pine M. N. Taylor (1981)

The Haunted Mansion and Pines

Merrill

Trees typically heighten and enhance their surroundings, and if their surroundings are sinister, they tend to take on a mysterious, spooky quality themselves. Such, at least, seems to be the case with the foreboding white pine by the front porch of the Scott mansion in Merrill, in Lincoln County, as well as the dark, thickly spaced pines planted around it.

The land the "Haunted House" now occupies once belonged to Native Americans. An early name for the area was Jenny Bull Falls, or, later, Jenny, after a beautiful maiden. It seems that well over a hundred years ago, when white lumbermen came north along the river, they were welcomed by a stately chief of the French-named Squiteo-eau-Sippi tribe. The beauty of Jenny, the chief's daughter, charmed the men. One version of the tale holds that Jenny became pregnant by one of the white men and felt it necessary to kill herself in the river. Another version relates that Jenny caught the flu when an epidemic swept the white lumbering community. In either case, her father blamed the white men for her death. He buried Jenny on the hill (on which the mansion was later to stand), and cursed any white man who would, in the future, dare to violate her grave.

The curse soon had a chance to work. Merrill was incorporated in 1883 and the first mayor, T. B. Scott, decided the next year to build a home on the top of the hill where Jenny's grave was. Even before the house was finished, he died suddenly. The next year his wife, who had continued the construction work, died, and in the same year the Scotts' son was killed by the home's architect, after an argument about whether or not the hill was cursed. In 1893 a Chicago millionaire bought the home. Six years later, he lost the home to a mortgage forecloser, went insane and died in an asylum. That same year, the man who now held the mortgage was stabbed to death. A land speculator who next bought the house disappeared soon after, never to be seen or heard from again. A woman was the next owner and she died within a year. From the time of her death until 1911, a caretaker managed the premises. Then he visited England, and returned on the *Titanic,* becoming one of the hundreds who perished. The next caretaker died of alcoholism. Finally, in 1919, the widowed husband of the woman who had last owned the property gave it to the City of Merrill.

In 1923 the city gave the land and the house to the Sisters of Mercy of the Holy Cross, apparently ending the curse.

Source: M. N. Taylor, Merrill

Chapter X

Arborphiles

Walter Scott in front of boxelder at Hickory Hill B-Wolfgang Hoffmann (1979)

Walter Scott's Hickory Hill

Madison

What is a weed? A plant whose virtues have not yet been discovered.

—Ralph Waldo Emerson
in *Fortunes of the Republic,* 1878

To some people the boxelder is a messy, overcompetitive weed tree species. But Walter Scott, whose fondness for Wisconsin trees was a life-long passion, found virtue even in this representative. In the dooryard of Hickory Hill House, the historic Madison home in which Walter and his wife, Trudi, lived for over 40 years, stood a remarkable 100-year-old boxelder. Its gnarled trunk had the character of an oak, its crown all the spread of a chestnut, and in height and stature it rivaled the pride of an elm.

When the state record book of big trees was begun in 1941, this tree was one of the first to be measured and entered as a state champion. During the years he served in the Wisconsin Department of Conservation (later to become the Department of Natural Resources), Walter was an enthusiastic participant in the big tree program. He served as an official record keeper for 28 years until his retirement in 1974. Even though his boxelder was displaced in the record book by larger contenders, he faithfully measured the tree on the first day of spring each year and recorded annual growth rates.

As one would expect, there are many other trees in Hickory Hill. In fact, it is a veritable arboretum with more than 100 species represented. One tree of special interest was an American spindle tree (wahoo) that was grown from a cutting taken from the state record specimen beside Aldo Leopold's shack in Sauk County.

Another dooryard tree that is probably as old as the house itself is a lilac. Years ago a coal truck broke off part of it, but it still thrives and blooms each spring.

The home was built outside the city limits of early-day Madison in 1842. John G. Hicks, who came from the East, was its first owner. It is now a designated Landmark Madison home and typifies the sturdy stone farmhouses of the time.

The location of Hickory Hill (its original acreage bordered the southwestern shore of Lake Mendota for one half mile) is certainly historic. In December 1974, school children studying the history of their home neighborhood solicited Walter's help in locating large oaks that might have "witnessed" the retreat of Black Hawk and his doomed followers in July 1832.

Walter went one step further and spent the Christmas and New Year's holidays conducting an extensive survey of the trees within a one-mile radius of his house. He measured and recorded more than 700 trees, then compared his observations with those made in 1835 by the original government surveyors. Not only did he locate the large and historic trees as requested—the 200-year-old oaks—but he also came to some interesting conclusions on the changes in vegetation that had occurred during the 140-year interval. These were published in an article in the Botanical Club of Wisconsin newsletter in April 1975.

As part of the Bicentennial celebration in 1976, Walter began a study of oaks within an 8-mile radius of the capitol, looking for oaks over 200 years old. He established a size minimum of 10 feet in circumference at 4½ feet above the ground for bur and white oaks, and, because they grow faster, 12 feet for black, red and pin oaks. Scott figured that all trees at least the minimum size would surely be at least 200 years old. Approximately 443 oaks in Madison met or exceeded this minimum size.

In 2001, R. Bruce Allison carried on the tradition of the Bicentennial Project. He attempted to relocate Walter Scott's original "bicentennial oaks" 25 years later, and was able to find about half of them. The others were either gone or couldn't be found because the city had grown, or because of changes to road names or locations. Those oaks that Allison was able to locate are now recorded using satellite-based Global Positioning System (GPS) coordinates to make these links to our past easier to find.

Walter Scott died in 1983. One of the two main scaffold limbs of the Hickory Hill dooryard boxelder fell shortly after Walter's death. The other main limb fell when his wife, Trudi, moved away.

Walter Ware Pines

Waushara County

A living monument of century-old pines still does honor to one of Wisconsin's first conservationists, Walter Macon Ware, who planted them near Hancock in Waushara County between 1867 and 1876 to commemorate the 1876 Centennial. They are located at Second Avenue and County Trunk 0, one-half mile west of the junction with KK, about four miles northwest of Hancock.

Today they remind us of what the state might have been like if more of Wisconsin's pioneers had been as interested in conservation and reforestation as they were in lumbering and exploitation.

Ware, who came from Hancock, New Hampshire, to Hancock, Wisconsin, in 1856 at the age of 21, took seven years to transplant the 1,876 pine seedlings to celebrate the nation's 100th birthday.

The pines themselves came from the nearby Leola Marsh. There, nourished through dry spells by the boggy terrain, grew giant white and Norway pines. Seedlings from these huge trees sprouted on whatever high ground was available, and it was these seedlings that Ware carefully dug up, preserving

Walter Ware Pines (c. 2003)

their root ball of moist, peaty soil, and transplanted by hand to the marginally fertile sands of central Wisconsin. Oxen, pulling a heavy stone boat, provided the transportation for the seedlings as Ware and his family provided the labor.

Some of the seedlings were set close together as a shelter belt against the winds, while the others were planted about eight feet apart. In spite of the change from marsh to sand, under Ware's care, virtually every transplanted seedling survived. Soon the plantation began to stand out like a green oasis, the only one of its kind in the state. Ware did not limit his forestation activities to pines. He also set out what became a fine stand of butternut trees destined to produce bountiful crops for generations.

When Ware's pine plantation was 12 years old he moved to Arkansas, but, as the years went on, he was recognized as having bestowed a fruitful legacy on Wisconsin, in contrast to many early settlers who left behind them slag heaps, abandoned mines and cut-over land. The Leola Marsh, drained, turned tinder-dry, and fire destroyed the giant ancestral pine trees as well as the younger siblings of the Ware pines.

The drought of the early 1930s proved the efficacy of tree shelter belts, by then advocated by the University of Wisconsin College of Agriculture experiment station at Hancock, which was studying ways of successfully farming the sandy soil. In the '40s Walter Ware was recognized by the Wisconsin Department of Agriculture. Their state bulletin proclaimed: "Would that more of the pioneers of this state had been as far-sighted and interested in planting trees as well as cutting them down. If there had been more of his type, our saw mills would still be humming today and ghost towns of the old logging days would be unknown."

Nevertheless, over the years, the pine plantation was mainly ignored by a succession of owners until, in April 1961, Floyd Foster sought the assistance of the district forester, Alex Katovich, in making a long-delayed harvest. Over 25,000 board feet of sawlogs were taken from the seven-acre plot, leaving 80,000 feet to grow and thrive.

Now in at least their 125th year, the pines in Ware's plantation have witnessed the flights of the once-multitudinous passenger pigeons, who nested in the great pines on the border of the Leola Marsh. They've seen the marsh drained and the fires come and the great pines die. They've seen the dust storms of the early '30s prove the effectiveness of pine plantings as shelter belts in the sand country. Planted laboriously and lovingly by hand, on days too rainy for farm work but ideally suited to the uprooting and transplanting of tender seedlings, they've seen the introduction of completely mechanized tree planting equipment that can handle 15,000 seedlings a day.

The Ware plantation trees, set out to commemorate the nation's Centennial, were still standing for the Bicentennial. They are a tribute to the energy and foresightedness of one of the state's pioneer settlers.

Sources: Alex Katovich, Waushara County
Barbara Vroman, Hancock
Tracy Salisbury, Green Bay

Donald Redman: Ripon's Tree Man

Before it lost its elms, Ripon, in Fond du Lac County, called itself the City of Trees. Fortunately, it was also the city of Donald Redman, a biology teacher and a man dedicated to tree planting. Over the years, he and his students planted a staggering total of more than 150,000 trees, bought with money they raised themselves through projects such as collecting aluminum cans and waste paper for recycling.

In 1968, Mr. Redman had his middle school students make a study of Ripon's trees, noting the percentage of elms: 32. The point of the study was to get city officials thinking in terms of replacement.

Mr. Redman's own private replacement program never faltered. For four or five weeks every fall semester, students in Mr. Redman's biology classes would walk around their neighborhood identifying trees. They located more than 50 different species. They also planted more than 300 trees on school property, their own and others. One of their first projects was planting maple trees in a nearby park.

What got Mr. Redman started on his tree planting crusade was needless destruction. One day, in a "road widening" project, the City of Ripon cut down 68 fine and healthy trees in the space of three blocks. These included two sugar maples in perfect condition that stood in front of Mr. Redman's home. So angry, so appalled, so incensed was he that he vowed it would never happen again. He ran for a seat on the City Council at the next opportunity, vowing to raise people's tree awareness. He won, and served for 12 years, during which time the city did adopt a tree-planting program.

On his own four acres within the city, Mr. Redman planted more than 3,500 trees over the years, mostly white and red pine, but also black walnut, apple, silver maple, mockernut hickory and basswood. He was particularly fond of conifers because he grew up near Green Lake in the sand country and as a small boy remembered being scared when the sand blew so hard it was impossible to see. At that time, in the 1930s, the planting of evergreens as shelter belts or windbreaks was a new technique that fascinated him.

Sources: Georgeann Penson, Ripon
Donald Redman, Ripon

Joyce Kilmer: The Man Who Wrote "Trees"

I think that I shall never see
A poem lovely as a tree.
A tree whose hungry mouth is prest
Against the earth's sweet flowing breast.

A tree that looks at God all day,
And lifts her leafy arms to pray;
A tree that may in Summer wear
A nest of robins in her hair.

Upon whose bosom snow has lain;
Who intimately lives with rain.
Poems are made by fools like me,
But only God can make a tree.

—Alfred Joyce Kilmer, 1913

Joyce Kilmer, the man who wrote "Trees," surely one of the best known and loved poems in the language, had a Wisconsin connection. He found the beautiful tree-shaded grounds of Campion College in Prairie du Chien much to his liking. In 1912, the busy editor, writer and lecturer began a correspondence with Father James Daly, a professor of English literature at Campion. It developed into a firm friendship. In 1913 the poem "Trees" appeared in the literary magazine *Poetry*, and was almost immediately acclaimed. But Kilmer did not make his first visit to the bluffside Catholic College at the confluence of the Wisconsin and Mississippi Rivers until 1916. He embraced it at once. According to his secretary, John Bunker: "Certainly there was no visiting place to which Kilmer looked forward with such pleasant anticipation and from which he returned so physically and spiritually refreshed as Campion."

Although Kilmer wrote 87 poems collected into three volumes, none achieved the fame of "Trees," a fame that saw eastern trees vying, through their human spokesmen, for the honor of having inspired it. Among the contenders are a white oak on the campus of Rutgers, his alma mater; a maple in Montague City. Massachusetts; oak trees surrounding his home in Mahwah, New Jersey; maple trees in Swanzey, New Hampshire; and a woods near Darien, Connecticut.

Kilmer first visited Campion to lecture on the topic "The War and the Poets" in April 1916. He later delivered the Campion College commencement address on June 15, 1917, while wearing the uniform of the New York National Guard, in which he had voluntarily enlisted. Soon after returning east, he requested a transfer to the infantry and on October 31, sailed for France.

Joyce Kilmer

The next year, on July 30, 1918, at the age of 31, he died in action, the first American man of letters to fall in "the war to end all wars." According to his correspondent and friend, Father Daly, he was "the most distinguished and widely regretted American to fall in battle during the Great War." He is buried near the field of battle a few miles from Seringes on the River Ourcq.

Friends and family dedicated the Kilmer Library at Campion in 1937. The poet's correspondence with Daly, other letters, personal effects, books by and about him, manuscripts and typescripts of poems, periodicals containing Kilmer's verse and essays, photographs and other memorabilia form the basis of the library. His widow, Aline Kilmer, said, "The Library at Campion is, I believe, of all memorials the one Joyce would have loved best."

When Campion closed its doors in 1978, the Kilmer collection went to Marquette University Memorial Library in Milwaukee.

Sources: Philip M. Runkel, Milwaukee
Griff Williams, Prairie du Chien

Rudy Lange Courtesy Louise L. Bauman (1931)

Rudy Lange: Delavan Tree Surgeon

Rudy Lange was one of Wisconsin's truly extraordinary tree surgeons. He was born in the family home in Delavan on April 1, 1907. His father, a teacher at the Wisconsin School for the Deaf, was also a nationally known arborist and was instrumental in having Arbor Day observed in Wisconsin. Rudy learned his tree-climbing and arboricultural skills from his father and when he was old enough he took over the business in Delavan. For over fifty years, until his death in 1981, he was a leader in the field of tree care. An extremely strong and energetic man, he set a hard pace to follow.

His friend and fellow naturalist Delavan postmaster W. Gordon Yadon wrote of some of the adventures for which Rudy was noted: "As a teenager, he swam alone across Delavan Lake; was a barnstorming stunt pilot; nearly perished in the 1936 fire at the Wisconsin School for the Deaf; as a scuba diver explored and chartered ship wrecks in Lake Michigan; encountered sharks and Moray eels off the coast of West Africa; was seriously stung by a Portuguese Man of War in the Caribbean; was nearly swept to sea by strong current off the Canary Islands; recovered the bodies of numerous Walworth County drowning victims; skied on treacherous slopes in the Swiss Alps; at age 66 felled two thugs who attacked him in a robbery attempt in Bogota, Columbia, received a deep knife wound in the leg which required 17 stitches and ten days later defied piranha and crocodiles to swim the Amazon River in Peru; at age 70 made an excursion to the Himalayan Mountains in Nepal where he stayed with Sherpa natives at 14,500 feet altitude near Mt. Everest; encountered anacondas in South America, cobras and tigers in India and handled and detonated tons of explosives during WW II and in his tree business." He died, ironically, in a home accident. As requested, his cremated remains were incorporated in the soil at the base of his favorite tree, an Adams flowering crabapple.

H. D. Thoreau: Wisconsin Tourist

Henry David Thoreau seldom ventured far from Concord. His most far-reaching excursion did, however, carry him across the state of Wisconsin. Thoreau and 17-year-old traveling companion Horace Mann, Jr., son of the famed educator, set out from Concord on May 11, 1861. They traveled overland to the Mississippi River, putting in at East Dubuque, then up to the St. Paul area, going as far west as Red Wing, Minnesota, before returning down the Mississippi and disembarking at Prairie du Chien on June 27. They traveled by train across southern Wisconsin, following the Wisconsin River, passing through Madison, the Kettle Moraine area, and finally arriving at Milwaukee, where they boarded a lake ferry to continue their return journey.

Thoreau's travel notebook is devoted largely to botanical observations with many references to trees. Thoreau held the white pine in special reverence, once writing that "it is as immortal as I am and perchance will go to as high a heaven, there to tower above me still." As his Mississippi riverboat passed La Crosse, the only observation of the region worthy enough to be penned in his notebook was "White Pine began half a dozen miles above La Crosse."

Another species that caught Thoreau's eye and imagination during the trip was the wild apple. His first observation of this species occurred on the train ride through Illinois to the Mississippi. He noted: "flowered apple-like trees." Then while staying at a boarding house on Lake Calhoun in Minnesota, he made a special effort to locate and identify a cluster of wild apples. No doubt, while traveling through Wisconsin he observed additional clusters of wild apples along the train route. These discoveries provided some thoughts for an essay entitled "Wild Apples," written upon his return to New England and shortly before his death in 1862. In this essay Thoreau described the fight of the wild apple for survival from browsing cows and the encroachment of civilization.

Henry David Thoreau Wisconsin Historical Society, WHi 10828

Index

References to photos are in italic type.

A

B

C

D

E

F

G

H

I

J

K

L

M

N

O

P

R

S

T

U

V

W

Y